Fluid Power Engineering

D. E. TURNBULL, Ph.D., B.Sc.(Eng), C.Eng., F.I.Mech.E., M.R.Ae.S., M.I.Prod.E., Mem.A.S.M.E.

Associate Professor of Mechanical Engineering, The University of Petroleum and Minerals, Dhahran, Saudi Arabia

With contributions by

J. K. Royle, Ph.D., M.Sc.Tech., C.Eng., F.I.Mech.E., F.R.Ae.S.
Professor of Mechanical Engineering, University of Sheffield

B. S. Nau, Ph.D., B.Sc., A.R.C.S.
Group Head, BHRA Fluid Engineering, Cranfield

London

Newnes-Butterworths

THE BUTTERWORTH GROUP

ENGLAND	Butterworth & Co (Publishers) Ltd London: 88 Kingsway, WC2B 6AB
AUSTRALIA	Butterworths Pty Ltd Sydney: 586 Pacific Highway, NSW 2067 Melbourne: 343 Little Collins Street, 3000 Brisbane: 240 Queen Street, 4000
CANADA	Butterworth & Co (Canada) Ltd Scarborough: 2265 Midland Avenue, Ontario M1P 4S1
NEW ZEALAND	Butterworths of New Zealand Ltd Wellington: 26–28 Waring Taylor Street, 1
SOUTH AFRICA	Butterworth & Co (South Africa) (Pty) Ltd Durban: 152–154 Gale Street

First published in 1976 by Newnes–Butterworth

ISBN 0 408 00199 2

Filmset and printed in Great Britain by Thomson Litho Ltd., East Kilbride, Scotland

Preface

This book has been written to provide a basis of a course on fluid power engineering by bringing together the theoretical and practical information available on the topics considered. In addition special attention has been given to providing numerous references so that practising engineers may find it a useful source of information on items of particular interest.

The field covered by the subject is now so large that it is difficult to set well-defined limits and, as a result, some readers may feel that certain areas have not received the attention they deserve. However, the aim has been to provide information on most of the main items commonly encountered in simple fluid power systems. Particular attention has been given to both analytical treatments and physical descriptions of the characteristics and design techniques associated with commonly used components.

In a few instances a very small amount of duplication was considered advisable to maintain conformity with current practices in terminology and notation in different but related fields.

I am very grateful to Professor J. K. Royle for providing the introductory chapter and for his many constructive comments and suggestions concerning the whole text. I also acknowledge the help of Dr. B. S. Nau for contributing the chapter on seals and tender my thanks to the Authorities of the University of Petroleum and Minerals and my present colleagues whose assistance proved invaluable. I am naturally indebted to many of my former colleagues and present friends for their past help and suggestions and in particular to Mr. G. T. Eynon, Mr. L. E. Prosser, Mr. N. A. Shute and Mr. G. Hibbert. Lastly I wish to dedicate this book to my wife, Meropi, whose encouragement and everlasting patience made its preparation possible.

D.E.T.

U.P.M., Dhahran, Saudi Arabia

Contents

Chapter 1

Introduction

The well-established topic of oil hydraulics owes its popularity to the ease with which large forces, torques, or powers can be generated, transmitted and controlled. This in turn leads to large torque/inertia ratios or force/mass ratios in equipment which is to be controlled.

Oil is a viscous fluid and, as in all calculations involving fluids, the answers to any problem are just as good as the assumptions that are made in the derivation and solution of the various equations that arise. It may therefore be surprising to find that some applications of the art of oil hydraulics do not require or even mention viscosity explicitly; in other applications viscosity may be all important. Again, oil compressibility may or may not be a dominant term in the analysis and the same is true for many other parameters.

1.1 Fundamental equations

If we start from the basic fundamental equations of fluid dynamics we encounter what appears to be the rather formidable Navier–Stokes equations.

The terms u, v, w, are the velocity components in the directions Ox, Oy and Oz as shown in *Figure 1.1* and we obtain equations such as:

$$\frac{\partial u}{\partial t}+u\frac{\partial u}{\partial x}+v\frac{\partial u}{\partial y}+w\frac{\partial u}{\partial z}=F_x-\frac{1}{\rho}\frac{\partial P}{\partial x}+\frac{\mu}{\rho}\left(\frac{\partial^2 u}{\partial x^2}+\frac{\partial^2 u}{\partial y^2}+\frac{\partial^2 u}{\partial z^2}\right)+$$
$$+\frac{1}{3}\frac{\mu}{\rho}\frac{\partial}{\partial x}\left(\frac{\partial u}{\partial x}+\frac{\partial v}{\partial y}+\frac{\partial w}{\partial z}\right) \qquad (1.1)$$

The equation simply states in mathematical form, that the net force in the Ox direction on a small volume of fluid gives rise to a rate of increase in the momentum in the Ox direction. Two similar equations arise which refer to the Oy and Oz directions, and the derivation of these equations may be found in most books of advanced fluid mechanics.

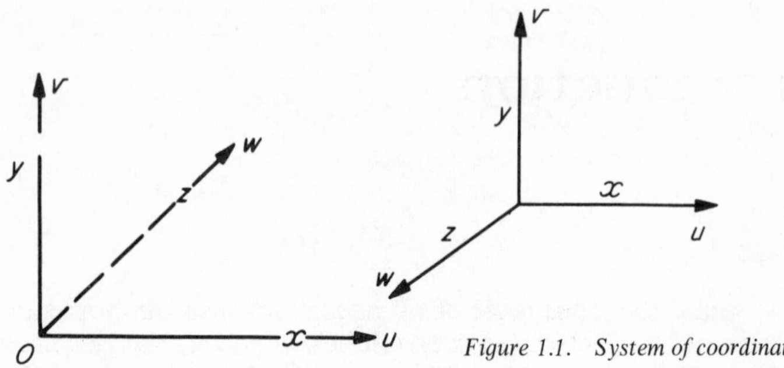

Figure 1.1. System of coordinates

The equations are very rarely, if ever, solved with all the component terms present; instead, the important terms only are considered and other terms are ignored as being unnecessary and unimportant in a particular application or range of applications. This procedure then leads to much simpler equations from which answers can be readily derived. Alternatively, one may make the simplifying assumptions and set up the simplified equations directly provided one remembers what has been omitted in making the assumptions.

1.1.1 Inertial terms

Thus, if one assumes a particular case where the only terms that really matter concern the pressure gradient and the inertial terms in the Ox direction we have from equation 1.1

$$u\frac{\partial u}{\partial x} = -\frac{1}{\rho}\frac{\partial P}{\partial x}$$

which is the Bernoulli equation generally written simply as

$$P + \tfrac{1}{2}\rho u^2 = \text{constant}.$$

The term $\partial u/\partial t$ in equation 1.1 is a temporal acceleration term. Thus, a steady state system may well include an acceleration $u(\partial u/\partial x)$ for example where the flow along a pipe changes due to change of area of the pipe along its length but this does not involve temporal change. If the flow rate along the pipe now changes with time we have the additional acceleration term $\partial u/\partial t$.

The expression F_x in equation 1.1 is referred to as a body force, i.e. a force which acts at a distance, as opposed to surface force (in the form of pressure or viscous stress). An example is the force due to gravity. If we rewrote equation 1.1 for the vertical direction and simply considered a static tank of fluid, we would have a vertical force $-\rho g$ acting downwards and the equation then reads:

$$+\rho g = -\frac{\partial P}{\partial y}$$

or

$$P = -\rho g y + \text{const.}$$

which simply expresses the vertical variation of static pressure in a tank.

Similarly we may have hydraulic equipment, e.g. a simple tank which has horizontal acceleration a, giving rise to horizontal pressure changes expressed by $\mathrm{d}P/\mathrm{d}x = \rho a$. In this accelerating system we have simply applied d'Alembert's principle and replaced the accelerating system by a static system with the inclusion of a body force.

1.1.2 Viscous terms

If we consider a viscous fluid flowing between two plane, stationary surfaces, which might be slightly inclined (*Figure 1.2*) we can assume, in the simplest case that the flow is only in the Ox direction. If

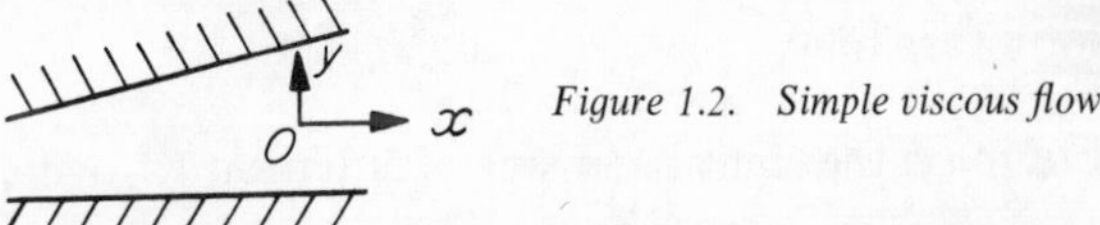

Figure 1.2. Simple viscous flow

the only terms that now really matter concern the pressure gradient and the viscous terms we have, again for equation 1.1.

$$O = -\frac{1}{\rho}\frac{\partial P}{\partial x} + \frac{\mu}{\rho}\frac{\partial^2 u}{\partial y^2}$$

or

$$\frac{\mathrm{d}P}{\mathrm{d}x} = \mu\frac{\mathrm{d}^2 u}{\mathrm{d}y^2} \quad \text{and} \quad \mu\frac{\partial u}{\partial y} \quad \text{is the viscous stress } \tau$$

If we further accept the experimental fact that fluids do not slip at solid boundaries i.e. $u = 0$, when $y = \pm(h/2)$, where h is the clearance the equation is readily integrated to give

$$u = \frac{1}{\mu}\frac{\mathrm{d}P}{\mathrm{d}x}\left(\frac{y^2}{2} - \frac{h^2}{8}\right)$$

which is the familiar parabolic velocity distribution. The flow rate in the Ox direction, per unit thickness (of Oz direction), is then simply

$$q_x = \int_{-h/2}^{h/2} u \, \mathrm{d}y$$

and this integrates to the form

$$q_x = \frac{-h^3}{12\mu}\frac{\mathrm{d}P}{\mathrm{d}x}$$

which is a basic equation relating to viscous flow in thin films, and forms a basic part of equation 45 (see Chapter 4).

We have seen that expressions such as $\mu(\mathrm{d}^2u/\mathrm{d}z^2)$ represent viscous terms where, for example the shear stress $\tau = \mu(\mathrm{d}u/\mathrm{d}z)$. Fluids which have a proportional relationship between shear stress and velocity gradient are termed 'newtonian fluids'. This book is concerned only with these fluids but it should be noted that the assumption is not strictly true for all fluids in all circumstances. This book is not primarily concerned with non-newtonian properties such as visco-elasticity, this thixotrophy etc. which, in certain fluids, can be pronounced, but some examples of such properties are given in *Figure 2.2* (see Chapter 2).

1.1.3 Compressibility

The last term on the right-hand side of equation 1.1 namely:

$$\frac{\partial}{\partial x}\left(\frac{\partial u}{\partial x} + \frac{\partial v}{\partial y} + \frac{\partial w}{\partial z}\right)$$

is a dilation term. With an incompressible fluid the term is

$$\frac{\partial u}{\partial x} + \frac{\partial v}{\partial y} + \frac{\partial w}{\partial z} = 0$$

and this is simply a mathematical statement of the fact that the volume rate of flow into a fixed control volume is precisely equal to the volume rate of flow out of the control volume for an incompressible fluid.

Although the complete term

$$\frac{\mu}{3\rho}\frac{\partial}{\partial x}\left(\frac{\partial u}{\partial x}+\frac{\partial v}{\partial y}+\frac{\partial w}{\partial z}\right)$$

is rarely of significance in calculations, it does not follow that *compressibility* can always be ignored.

In fact, all fluids, including oil, exhibit compressibility. If we have a volume of fluid V, then an increase in pressure of $\mathrm{d}P$ will change the volume of fluid by an amount $\mathrm{d}V$ so as to reduce the volume. We define the bulk modulus $K' = -V(\mathrm{d}P/\mathrm{d}V)$ i.e. as a positive number and this term may be either unimportant in a calculation or may alternatively be of very great importance. The latter case arises when we have a long column of oil subjected to pressure changes and supporting an inertial load.

Just as a steel column supporting a load will vibrate due to the interchange of strain energy in the column and the kinetic energy of the supported mass, so an oil column and associated mass will exhibit a natural frequency of oscillation and it is found that oil is approximately one hundred and fifty times more compliant than steel. Thus whenever large moving masses are involved in hydraulic equipment, compressibility invariably becomes a point of great importance. As is shown later (see section 2.2.1, Chapter 2), when air is entrained in the oil, the stiffness of the oil column can be reduced to very low values.

1.1.4 Dimensionless terms

It is fairly obvious that in many applications of fluid power we would be straining the assumptions if we were to completely ignore say the inertial terms in equation 1.1 and concentrate only on the viscous terms, or vice versa. It is therefore important to study the magnitude of the various terms in equation 1.1.

Consider a system involving hydraulic components in which we may forget about body forces such as gravity, where compressibility is assumed to be unimportant, and where the fluid flow is essentially one dimensional, e.g. along a pipe. Equation 1.1 will then read

$$\frac{\partial u}{\partial t}+u\frac{\partial u}{\partial x}=-\frac{1}{\rho}\frac{\partial P}{\partial x}+\frac{\mu}{\rho}\left(\frac{\partial^2 u}{\partial y^2}+\frac{\partial^2 u}{\partial z^2}\right)$$

If a representative length L, and a representative velocity U are chosen, all velocities and lengths may be expressed as fractions of U or L so that, for example $u = u'U$; $x = x'L$ where u' and x' are dimensionless. Further, the system may be oscillatory and have a natural frequency of oscillation denoted by f, where f has dimensions of 1/time i.e. $1/T$.

The modified form of equation 1.1 now reads

$$\left(\frac{fL}{U}\right)\frac{\partial u'}{\partial t'}+u'\frac{\partial u'}{\partial x'}=-\frac{\partial^{(P'\rho U2)}}{\partial x'}+\frac{\mu}{\rho UL}\left(\frac{\partial^2 u'}{\partial y'^2}+\frac{\partial^2 u'}{\partial z'^2}\right) \tag{1.2}$$

In this dimensionless form the term $(\rho UL)/\mu$ is dimensionless and is the Reynolds number (Re). Physically, it is a measure of the ratio of inertial terms to the viscous terms. A low Reynolds number implies high viscosity, or very low velocity, or very small size.

The term $(fL)/U$ is the Strouhal number and again is the dimensionless ratio of transport time (the time for a fluid moving with velocity U to cover a distance L) to the periodic time of oscillation as measured by $1/f$. The dimensionless term $1/\rho U^2$ is associated with the name of Euler.

The Strouhal number is rarely important in hydraulic calculations—although exceptions can certainly be found—and it is introduced only to underline the statement that it is important to understand the nature and magnitude of assumptions implicit in all calculations involving fluid flow.

If we ignored the temporal term $\partial u'/\partial t'$ in equation 1.2, we could immediately compare the early equation from which the Bernoulli equation followed, i.e. $u(\partial u/\partial x) = -1/\rho(\partial P/\partial x)$ and we see that this simple equation is not therefore likely to be true unless the Reynolds number is very high.

All coefficients which are used in fluid power engineering depend on the Reynolds number—to varying degrees of importance. If we had been concerned with gas flow, we would have found that the dimensionless Mach number was important, and if heat transfer was of great significance, again we have appropriate dimensionless groups to express the ratios of relevant terms.

1.2 Examples of hydraulic calculations

Oil hydraulics is concerned with such terms as the flow through pipes and valves and leakage flow through small clearances.

As the fluid in contact with a solid surface is always at rest, the mathematical complications mean that very few practical problems can be solved precisely. This is in keeping with the whole of fluid mechanics where precise theoretical solutions are few and real problems are hedged, in varying degrees, by coefficients which, in our case, are dependent on the Reynolds number and the geometrical shape of the hardware.

1.2.1 Pipe flow

If we consider a pipe through which oil is pumped at a constant rate (*Figure 1.3*), a pressure drop occurs due to the shear forces acting

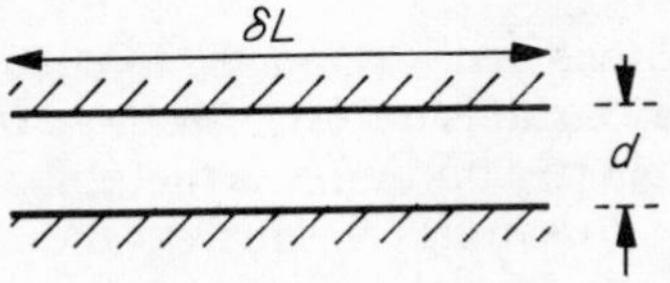

Figure 1.3. Pipe flow

on the fluid at the wall. If the pressure drop over a length of pipe δL is ΔP then we have

$$\tau \pi d\, \delta L = \Delta P \frac{\pi}{4} d^2$$

The magnitude of the shear stress τ depends on the nature of the flow pattern, which in turn depends on the Reynolds number (*Figure 1.4*). In a pipe, and at say a distance of 100 pipe diameters or so from the inlet, the flow is either laminar or turbulent and the

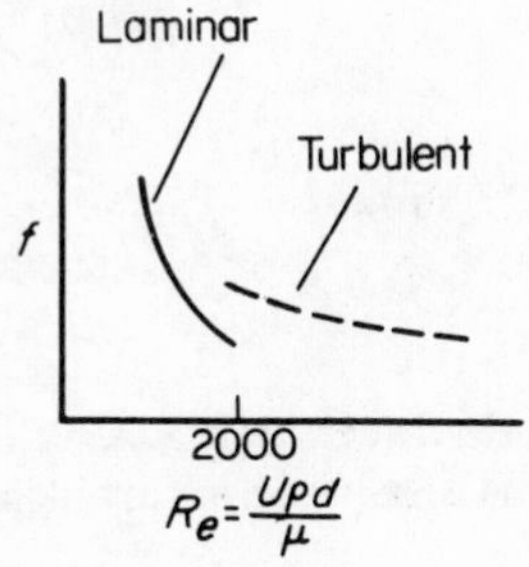

Figure 1.4. Variation of f with Reynolds number

change from laminar flow occurs at a Reynolds number of about 2000, based on the diameter of the pipe, i.e. $U\rho d/\mu$ where U is the mean velocity defined by $Q = UA$ (where A is the area of the pipe).

If we express τ as a fraction of the dynamic pressure $\frac{1}{2}\rho U^2$ and define $f = \tau/(\frac{1}{2}\rho U^2)$ then f is non-dimensional, and

$$\Delta P = \frac{4\rho f U^2}{2d} \delta L$$

or in terms of head loss, h_f

$$h_f = \frac{\Delta P}{\rho g} \quad \text{and} \quad \delta h_f = \frac{4fU^2}{2gd} \delta L$$

For laminar, i.e. viscous type of flow, the velocity distribution across the pipe is again parabolic, as we saw in the earlier illustration. For this type of flow in a pipe, we quickly find that $f = 16/\mathrm{Re}$. For turbulent flow we must ultimately refer to experiment and measure the flow rate and pressure drop to decide a suitable empirical relationship between f and Re.

Further losses in pipes are concerned with bends, T-junctions etc. In general, these losses are well established for various geometrical configurations and a reasonable rule of thumb, is that a sharp right angle bend causes a pressure loss roughly equal to the dynamic pressure $\frac{1}{2}\rho U^2$.

1.2.2 Orifice flow

The approach to orifice type flow is first to apply the Bernoulli equation as a guide to the solution of the calculation of flow rate. Essentially the pressure head is converted to velocity head through the orifice and the jet then separates from the solid boundaries.

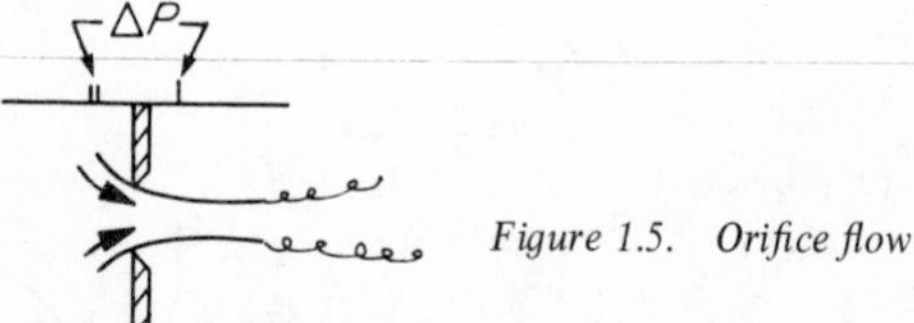

Figure 1.5. Orifice flow

Downstream of the orifice, the jet entrains fluid (certainly when the jet is turbulent); the kinetic energy of the jet is dissipated in turbulent mixing and eventually appears as heat.

If we ignore the kinetic energy upstream of the orifice we can equate the pressure loss ΔP to the dynamic pressure of the jet according to the simple Bernoulli equation.

$$\Delta P = \tfrac{1}{2}\rho U_t^2$$

where U_t is the theoretical velocity of the jet.

In fact the actual velocity of the jet is a little less than this and we therefore include a velocity coefficient C_v such that the actual velocity of the jet $U = C_v U_t$ and C_v has a value of about 0.97.

At first sight one might expect the diameter of the jet to be equal to the diameter of the orifice. However, with a sharp edged orifice, as opposed to a shaped nozzle, the fluid approaching the orifice has a considerable momentum directed radially inwards. This momentum cannot be instantaneously destroyed and the jet continues to contract until the pressure becomes sensibly constant across the jet at the vena contracta.

The diminution in jet size is accommodated in a coefficient of contraction C_c which essentially relates jet area to orifice area and has a value of about 0.62 over much of the range of higher Reynolds numbers. The two coefficients C_c and C_v may then be lumped together to give the discharge coefficient C_d which has a value of about 0.6 over the range of higher Reynolds numbers.

It may be remarked that when a theoretical result requires a correction factor of the order 0.6 to give a real practical result, then there is something wrong with the theory. This is so in the present case where momentum flow at right angles to the axis of flow was completely ignored by the simple one-dimensional approach.

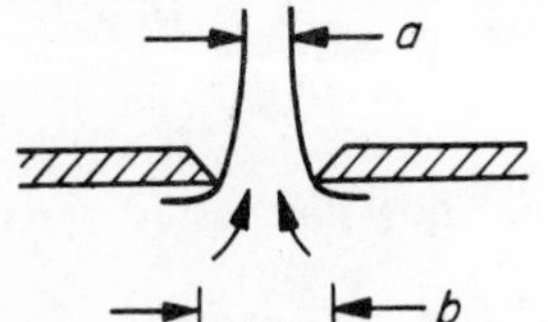

Figure 1.6. Flow through slit

In fact one can calculate, for a slit, i.e. a two-dimensional orifice in which the fluid again separates to form a jet (*Figure 1.6*), that the theoretical contraction of the jet, given by b/a, is $\pi/(\pi+2)$ or 0.611. This calculation assumes an inviscous fluid (i.e. in which viscosity is completely ignored) and it is included here to underline the point that a coefficient of discharge relating reality with simplified theory, even of the order 0.6 has a rational explanation when the 'correct' theory is used.

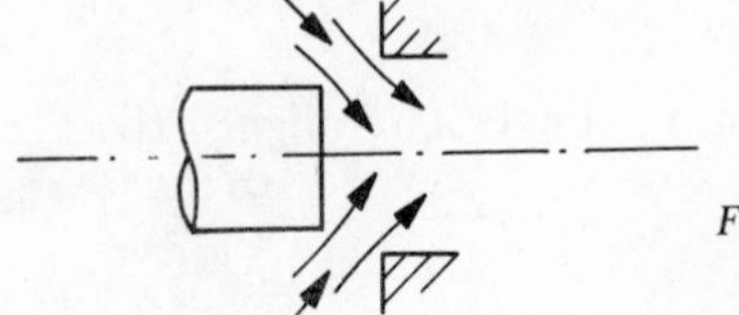

Figure 1.7. Flow through piston port

For the case of a two-dimensional orifice formed by a piston and cylinder with a circumferential inlet port (*Figure 1.7*), the same sort of inviscid analysis can be used to predict the angle and momentum in the jet. Further details of this appear in Chapter 6.

However, it can happen that other physical effects arise, which must be explained rationally.

In *Figure 1.8* the jet has reattached to the face of the piston. This *can* occur (but certainly does not always occur) and is due to the turbulent entrainment of fluid which produces a low pressure region

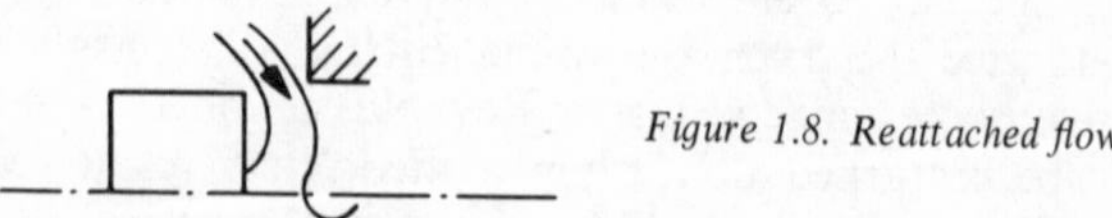

Figure 1.8. Reattached flow

and so bends the jet until reattachment occurs again further details of this are given in Chapter 6. It is frequently referred to as the Coanda effect and is very important in the study of fluidics.

1.2.3 Viscous flow

The basic equation for viscous leakage flow has been derived in the form

$$q_x = -\frac{h^3}{12\mu}\frac{dP}{dx}$$

for simple one-dimensional flow.

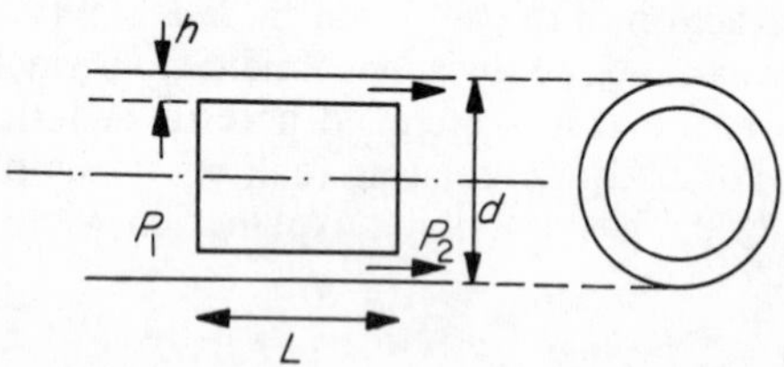

Figure 1.9. Viscous flow past a piston

If we consider say a piston in a cylinder one might expect the total leakage flow to be easily calculable. The 'width' of the leakage track is πd, the term

$$\frac{dP}{dx} = -\frac{P_1 - P_2}{L}$$

and the total leakage flow is then given by

$$Q = \frac{\pi d h^3}{12\mu}\frac{P_1 - P_2}{L}$$

Mathematically speaking, this expression is correct and highly accurate providing our basic assumptions are correct. In effect we have assumed that

1. We know the value of the radial clearance h accurately.
2. We assume that the piston remains concentric in the cylinder.

In real calculations both of these assumptions are usually highly suspect. If one erroneously assumes a small clearance to be uniform and of magnitude 1 unit, when in fact it is really 2 units, the error is to predict a leakage flow only 1/8 of the true value. As clearances are reduced and tolerances tightened, the sensitivity of leakage flow calculations with the cube of clearance is pronounced. In this case a poor result does not imply faults in the theory. The theory is remarkably accurate and what is generally wrong is the measurement.

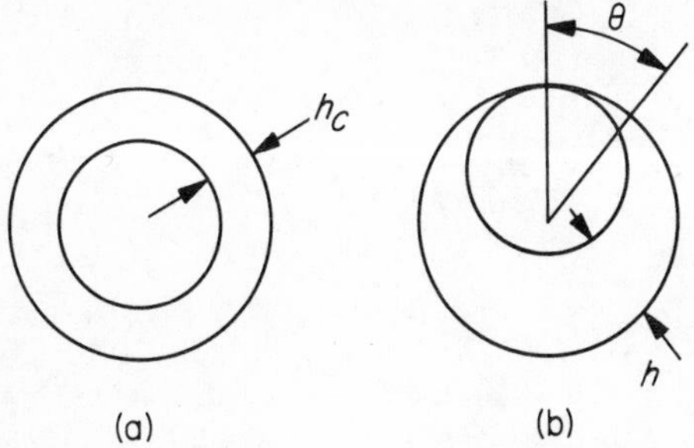

Figure 1.10. Concentric and eccentric pistons $h = h_c(1 - \cos\theta)$

Again the same sort of error arises if we challenge the second assumption, namely that the piston is concentric. The point is illustrated in *Figure 1.10* which shows the extreme cases of (a) concentric and (b) touching the cylinder wall so that the clearance varies from the defined radial clearance h_c according to $h = h_c(1 - \cos\theta)$. Under these conditions the ratio of the leakage flows in the two cases, with the same pressure gradient is given by

$$\frac{Q_b}{Q_a} = \frac{\displaystyle\int_0^{2\pi} h_c^3(1-\cos\theta)^3\,\mathrm{d}\theta}{\displaystyle\int_0^{2\pi} h_c^3\,\mathrm{d}\theta} = 2.5$$

Thus, on good theoretical grounds, the leakage flow can vary by a factor of 2.5 by concentricity arguments alone and it is a fact that concentricity, in most applications, is not known.

In the simple examples of some of the matters which arise in hydraulic power engineering given above, one must again stress the fact that the answers are just as good as the assumptions. Generally speaking, the physical phenomena of oil hydraulics are well understood although it is sometimes difficult to be precise about their magnitudes.

Chapter 2

Fluids for Hydraulic Systems

2.1 Introduction

At the beginning of this century several designs of variable speed hydraulic gears were being developed and, at the same time, oils rather than water began to be used as the driving medium in these machines. Since that time an extremely wide range of hydraulic oils has been developed specifically for use in hydraulic machinery and considerable advances were made in this field during and immediately after the second world war. In more recent years an entirely new aspect has received considerable attention, where attempts have been made to produce non-flammable fluids suitable for high speed aircraft and mining applications.

Water, because of its often undesirable properties, is now no longer regarded as a suitable medium for use in general hydraulic systems and although water-oil emulsions and water-glycol mixtures are used for special purposes, water alone can usually be disregarded. One of the few exceptions to this is the case of the London Hydraulic Power Company, which supplies 7000 h.p. to 180 miles of water pressure mains. Power from these mains drives the bascules of Tower Bridge, the revolving stage of the London Coliseum and, at one time, the front lift at the Institution of Mechanical Engineers. Its low vapour pressure and subsequently high evaporation rate, relatively high freezing point, low viscosity, corrosive action on certain metals and poor lubricating qualities make it one of the less attractive fluids to use in hydraulic systems. As a result is is almost invariably ignored when a suitable fluid is being selected.

Air is another freely available fluid the possible use of which at one time received considerable attention and, although numerous pneumatic servos and systems were developed in the decade

Table 2.1 RELATIVE MERITS OF SYSTEM FLUIDS

Fluid	*Advantages*	*Disadvantages*
Water	Generally available, fire resistant (Relative cost 0)	Poor lubrication, low viscosity, high vapour pressure and freezing point. Max temp 55 °C.
Air	As water, but has very high max working temp.	Poor lubrication, low viscosity, dangerous if a sudden burst, invisible leaks, low volumetric locking, high compressibility.
Hydraulic oils (a) Mineral based (b) Castor based	Good lubrication, high viscosity. (Relative cost 1.0)	Viscosity drops rapidly with rising temperature and to a lesser extent with falling pressure. Sometimes high air content. Max temp 55 °C.
Silicate esters (a) Polysiloxanes (b) Disiloxanes	Good for large temperature ranges. Good oxidation and thermal stability. Max temp 205 °C (but see disadvantages and section 2.5).	*Rapid* hydrolysis, attack metals at *high temperatures.*
Non-flammable fluids		
(a) Water glycols	Fair to good lubrication, very high viscosity index, high autogenous ignition temperature. (Relative cost 4.0)	High vapour pressure, attacks paints, zinc and cadmium. Max temp 55 °C.
(b) Water-oil emulsions	Fair to good lubrication, high viscosity index. (Relative cost 1.5)	High vapour pressure. Max temp 55 °C.
(c) Phosphate esters	Good lubrication, low vapour pressure, high autogenous ignition temperature, stable. Max temp 150 °C (Relative cost 5.0)	Low viscosity index, attacks synthetic rubbers, paints and aluminium, caution required in use, high cost. (Toxic vapour).
(d) Silicones	High viscosity index. Max temp 260 °C. (Relative cost 30)	Poor lubrication, sealing difficulties, low bulk modulus.
(e) Halogenated hydrocarbons (i) Chlorinated (ii) Fluorinated	Very good at high temperatures, good lubrication. Max temp 260 °C. (Relative cost 2.0 to 200)	Expensive, low viscosity index, high density. Attack paint. Toxic.

following the second world war, their popularity has waned considerably. Pneumatic systems are now generally only used in relatively low pressure and low power systems. This is due to several factors, including the high compressibility of the fluid (and therefore its lack of volumetric locking) and its poor lubricating properties. In addition, compressed air has a rather explosive-like nature if a component suddenly fails and this increases the hazards of its use. Although air is naturally very cheap and no fire risk accompanies its use, it has never been universally accepted in all fields of the fluid power industry.

Table 2.1 lists the main types of fluids which have been considered and used in systems together with their major advantages and disadvantages. Factors which influence the selection of a fluid for any particular duty will now be considered.

2.2 Choice of fluid

So many properties influence the choice of a fluid for a specific system that it is very doubtful if any one particular fluid can ever be said to be the best from every point of view. Some of these properties are extremely difficult to define and it is only recently that any attempt has been made to put their measurement on some form of scientific basis. Part of the difficulty is due to the fact that the behaviour of the fluid is very often a function of both the system in which it is to be used and the manner in which it is used, so that although a satisfactory performance may be obtained during the initial development work on the system, there is no guarantee that an acceptable performance will always be obtained in service in the field.

For example, consider a pump unit which is made as a general purpose fluid power supply. In one application it may be required to operate in the tropics, in another under semi-arctic conditions, in another at the coal face of a mine and in yet another near furnaces in a steel works. Each application will have different operating conditions in terms of ambient temperature, relative humidity and both the level and type of atmospheric contaminant will vary. Under tropical conditions it is probable that a continuous cooling system will be required whilst for semi-arctic conditions it is possible that a heating coil will be used in the oil reservoir at least before the unit is started.

The relative humidity may vary widely during each day resulting in a considerable amount of condensation so that the water content of the fluid would have to be examined frequently. The ability of the system to stand up to contamination will be severely tested in the mining application and again in the steel works. In the former, the contaminant will probably be of a mainly non-ferrous nature whereas in the latter it will almost certainly contain a large proportion of magnetic and acidic-type particles. This illustration shows only a few of the conditions to which fluids may be subjected and details of the more important properties influencing the selection of fluid are described in the next section.

2.2.1 General fluid properties

Many, if not all, of the following properties of a hydraulic fluid are usually examined before it is generally accepted as suitable for use in hydraulic systems:

1. Acidity.
2. Air content and affinity.
3. Aniline point.
4. Availability and cost.
5. Bulk modulus: Isothermal secant and tangent values; Isentropic secant and tangent values; Sonic value.
6. Corrosion resistance.
7. Emulsification resistance.
8. Foaming resistance.
9. Lubricating properties: Slip-stick, boundary, hydrodynamic.
10. Oxidation resistance (chemical stability).
11. Pour point.
12. Seal compatibility (see also aniline point).
13. Specific gravity.
14. Stability.
15. Storage life.
16. Toxicity.
17. Viscosity, absolute value.
18. Pressure and temperature coefficients.
19. Shear break down resistance.
20. Viscosity index.

Tests for the measurement of many of these properties (referred to later as IP numbers) are described in an Institute of Petroleum publication.[1] A wide range of other tests[2,3] is often used by engineers on site and for some of these no absolute standards exist. A summary of the more general tests, together with their meanings, is given in the following paragraphs.

1. *Acidity.* This test measures the total of the combined organic and inorganic acidity of the fluid and this quantity is measured in terms of the number of milligrams of potassium hydroxide that is required to neutralise completely one gramme of the fluid.

If the fluid is initially alkaline then the number of milligrams of the hydroxide required to neutralise the quantity of acid necessary to neutralise one gramme of the original fluid is determined. The result of this test is quoted as a *neutralisation number*, which is the actual number of milligrams of the hydroxide required. The Institute of Petroleum test reference number is IP 1/60, although an alternative method, IP 139/59 may also be used and has a tendency to give slightly higher values.

2. *Air content and affinity.* The amount of dissolved air present in most mineral based hydraulic fluids under atmospheric conditions is generally much greater than that for water and is usually between 8 and 16% by volume. As distinct from dissolved air a badly designed system can 'aerate' the fluid such that there is a considerable volume of free air (froth) trapped in the fluid and Reference 4 gives details of tests concerning this condition. No IP test is used for measuring air content but the foaming test described under (8) on p. 19 should be noted. Dissolved air does not generally affect the properties of the fluid but free air naturally has a great effect on system performance, particularly where the fluid bulk modulus is important. Since air rapidly dissolves in high pressure oil, measurements of free air effects at high pressure are difficult and no standards, etc., exist.

3. *Aniline point.* The minimum temperature at which the oil is completely miscible with an equal volume of aniline is defined as the aniline point. It gives an indication of the aromatic naphthenic and paraffinic contents of the oil, and generally its value increases with increasing molecular weight and decreasing quantities of aromatic and naphthenic constituents.

High aniline points are a characteristic of highly refined paraffinic oils which are relatively chemically stable and have a high viscosity index. A low aniline point indicates a high aromatic content which would produce a considerable swelling effect on rubber sealing materials. Although aniline point is not a complete guide to seal selection it does provide a useful indication, and if the fluid in a system is changed to another with a considerably different aniline point then it is unlikely that the original seals will be satisfactory.

4. *Availability and cost.* If world wide markets are being considered it is naturally essential that the fluid chosen and recommended for

use in a given system should be available in all parts of the world. In addition, there should be adequate supplies so that a large system storage tank does not have to be filled from a series of one gallon containers. Some of the larger oil companies operate world wide distribution services but care should be taken to avoid one which occasionally changes the composition of its products without giving notice of such changes.

From the nature of its origin one cannot expect that the basic constituents of a mineral oil will remain absolutely constant but it has sometimes proved extremely costly when the amount or number of additives in a 'standard' oil has been suddenly altered without any notice being given. It follows that occasional changes are necessary in order to incorporate the latest research and development results, but sometimes the value of these alterations is clearly not appreciated by a user who, having topped-up or refilled a system with the supposedly same fluid, finds, to his cost and inconvenience, that the amount of one or more of the additives has been altered, thereby making the oil completely unsuitable for the specific application. Some companies are naturally reluctant to disclose the complete analysis of their products but, nevertheless, most fluid users feel that when a change in composition has taken place notice of it should be given, even if it is only in relative rather than absolute terms.

The initial cost of a fluid is sometimes important if large quantities are required but generally the higher priced fluids contain considerably more additives so that their longer life and better operating qualities more than offset their greater initial cost.

5. *Bulk modulus.* There are five different but commonly accepted values of the bulk modulus of an oil, these being

(a) Isothermal secant.
(b) Isothermal tangent.
(c) Isentropic secant.
(d) Isentropic tangent.
(e) Sonic.

The secant value is a form of average value and is proportional to the slope of the line connecting a point on the pressure volume curve to the origin. The tangent value is proportional to the actual slope of the curve at any point and is generally of more practical value. The sonic bulk modulus is a name given to the isentropic bulk modulus when it is derived acoustically by measurement of the sonic velocity. Shute gives further details of these quantities and of methods of measuring them.[4]

The effect of free air on the value of the bulk modulus is considerable and an analysis of the problem has been derived by Hayward[5] who shows that the effective bulk modulus, K_m^1, of an air/oil mixture is given by

$$K_m^1 = \left\{\frac{(V_f/V_a)+(P_0/p)}{(V_f/V_a)+(P_0\,K'/p^2)}\right\}K^1$$

where V_a and V_f are the volumes of the air and oil at atmospheric pressure, p is the test pressure, P_0 is atmospheric pressure and K^1 is the bulk modulus of the air free oil. Other aspects of the characteristics of air and oil mixtures have been studied in detail by Khoklov.[6]

6. *Corrosion resistance.* So many types of metals are used in the construction of systems and their components that no universal test is available to measure the behaviour of oils in contact with each type of metal. There are two standard tests for steel and copper, IP 135/61 and IP 154/59. In the former a BS 970:1955—EN3B steel specimen is initially immersed in the fluid and then stirred in a sample of synthetic sea water for 24 hours. After this period, the specimen is examined for rust and the amount present is classified into one of four groups. The first, for satisfactory passing of the test, is no visible rust and the last is severe rusting, in which rust covers more than 5% of the surface. The copper corrosion test employs a polished, flat strip of copper which is placed in the fluid at a temprature of 100 °C for three hours. It is examined for tarnishing and corrosion after this period.

It should be noted that corrosion and oxidation properties are closely linked and can be assessed by the results of the acidity test described above.

7. *Emulsification.* Unless an oil separates rapidly from water there is a danger of it forming a stable water-oil emulsion and these can be either *thin slimy* or *thick pasty* substances both of which interfere with the working of a system. No standard test is at present used but attention should be paid to this most undesirable property.

8. *Foaming.* In this test air is blown at a constant rate through the sample of fluid, which is maintained at a temperature of 24 °C. The volume of the foam immediately after and also ten minutes after the air flow is stopped is measured. The test is repeated with a second oil sample at 93.4 °C (200 °F) after which it is stirred to collapse the foam, cooled to 24 °C and the first test repeated. The volumes of foam immediately after the air is cut off and at the end of the

ten minute settling periods represent the foaming tendency and stability respectively. Small amounts of anti-foaming additives can have a marked effect on the behaviour of an oil.

9. *Lubrication properties.* There are numerous bench tests designed to measure the lubrication properties of fluids but each gives distinct results which are mainly a property of the test apparatus and generally they bear little relationship to the behaviour of the fluid in an actual system. They include the Four Ball test, the Falex test and the Almen, Timken and S.A.E. tests. The most common one is certainly the 'Four Ball' test but despite a considerable amount of effort spent in trying to correlate the results of this test with practical results, to date, no outstanding success has been achieved.

So many types of lubrication conditions occur in the operation of most hydraulic units that it is impossible to describe them all, but three, generally accepted types are:

Stick-slip motion, which corresponds to solid friction conditions.
Boundary lubrication, which is very much a function of the surface finish of the bearing elements and the fluid used, and
Hydrodynamic lubrication where the bearing surfaces are separated by a complete fluid film (see Chapter 3, *Figure 3.1*).

10. *Oxidation.* Numerous oxidation tests exist and the results are rather like those for lubrication insofar as they are difficult to relate to practical conditions. However, IP 56/60; IP 114/56; and IP 157/56 should be noted together with a proposed rotary bomb test. In the first two the oil sample is subjected to a stream of air at 150 °C for 45 hours and at 95 °C for 90 hours respectively in the presence of a copper catalyst. At the end of the test the amount of sludge formed and the neutralisation number (see (1) above) are noted.

For IP 157/56, a 300 ml sample is held at a temperature of 95 °C in the presence of water, oxygen and an iron-copper catalyst. The neutralisation number is noted before, during and after the test, which is stopped when the neutralisation number reaches a previously selected value, generally between 0.25 and 2.0, or after a period of 1000 hours. This test is generally used for fluids having an oxidation inhibitor. The rotary bomb test is one in which a sample of the fluid is subjected to an oxygen pressure inside a vessel which is rotated at a fixed speed until the pressure falls by a predetermined amount. The time taken for this fall in pressure to occur is noted. Initially the pressure rises above the value to which the vessel was originally charged.

Other factors which accelerate oxidation are water and particle contamination and the products of corrosion. Since these may be present in varying degrees it is not possible to lay down a set life for an oil which may vary from a few months to several years and yet it is obviously important for a user to know when it is necessary to change the oil in order to avoid sludge forming in the system.

Oxidation is indicated by an increase in the total acidity of the oil and this can be detected before the sludge deposits become significant. However, hydraulic oils contain anti-oxidant additives whose function is to delay the onset of oxidation and until these are exhausted only small increases in acidity will occur. Towards the end of a fluid's useful life the acidity will start to increase at a much

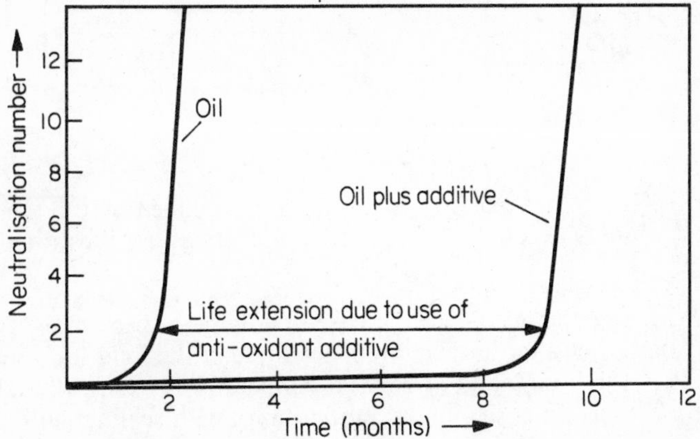

Figure 2.1. Variation of neutralisation number with time for an oil with and without an anti-oxidant additive

higher rate than it does when the fluid is first used and it is then that the fluid should be changed. To detect this change samples must be tested at regular and frequent intervals. Alternatively, a neutralisation number of about 2 may be taken as an indication that the rate of oxidation is in the region where it will increase rapidly as is shown in *Figure 2.1*.

11. *Pour point.* The pour point of an oil is the temperature which is 5 °F above that at which the oil just fails to flow when cooled under prescribed conditions. It is useful for estimating the approximate temperature at which the fluid's viscosity properties will become non-Newtonian. This term is used to describe a fluid for which the ratio, shear stress/shear rate is not constant but varies

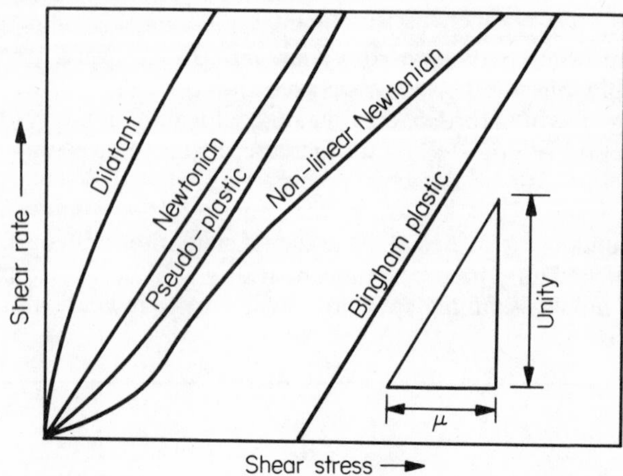

Figure 2.2. Relationship between shear stress and rate for Newtonian and non-Newtonian fluids

with the shear rate. Such types of fluids are defined by the type of relationship between the shear rate and stress as illustrated in *Figure 2.2.*

12. *Seal compatibility.* The increase in volume or percentage swell of an elastomer after immersion in a sample of the fluid for a given time (e.g. 168 hours) and at a given temperature is generally taken to give a measure of the fluids compatibility with sealing materials. Mineral based fluids are generally compatible with most synthetic rubbers and natural rubbers are often employed for applications requiring castor based fluids. Further details concerning sealing materials are given in section 4.3 of Chapter 4.

13. *Specific gravity.* This property is generally of little importance in hydraulic systems except where long pump suction pipe lengths are used. It is then essential to ensure that cavitation does not occur.

14. *Stability.* It is naturally important for a fluid to be chemically stable and not only be oxidation resistant but also to have no tendency to crystallise, gel or solidify. Low temperatures often accentuate these tendencies and one test is to maintain a sample of fluid at a low temperature (e.g. −54 °C) for a minimum period of 72 hours. After this time it is examined for turbidity and the other undesirable properties mentioned above.

A suitable standard of turbidity for reference purposes may be prepared from 125 ml of distilled water, 25 ml of 0.5 normal sulphuric acid and 25 ml of a 0.0032 molar solution of barium chloride. After thorough mixing the mixture is rendered alkaline by adding 25 ml of normal sodium hydroxide and made up to 250 ml with distilled water. A suitable quantity of red dye may be added to match the colour of the specimen tested. The solution must be used within 30 minutes of its preparation.

15. *Storage life.* A good storage life is desirable and the fluid should not change in chemical composition or separate out into its constituents during storage. It is usually recommended that drums of fluid be stored at a temperature between −30 °C and 30 °C and that the drums be rolled at periods not exceeding three months. If a lower temperature is experienced the fluid should be 'thawed out' at a temperature of 20 °C for between 24 to 48 hours.

Water absorption is a potential hazard and the drums should always be covered or at least stored horizontally so that water cannot collect on the top of the drum.

16. *Toxicity.* Some fluids, in particular the non-flammable types and to a lesser extent castor based fluids, may be toxic and the use of barrier creams may be mandatory. Some of the former may emit poisonous vapours, particularly if they are near to hot surfaces or flames and others may prove harmful to the skin.

Various dermatitic diseases can be caused by repeated contact even with mineral based fluids and care should be exercised at all times when they are handled.

17. *Viscosity.* This topic covers such a wide field that it has been broken down in into three portions covering the more important aspects.

Viscosity values are normally quoted at 37.8 °C and 98.9 °C (100 °F and 210 °F) and a straight line may be drawn through these points when they are plotted on a specially constructed chart (see *Figure 2.3*). These charts should only be used for fluids having Newtonian characteristics and hence values of viscosity should only be extrapolated to within approximately −9.4 °C (15 °F) of the pour point.

Numerous methods of both measuring and quoting viscosity values exist and this is perhaps a field where some International standardisation is urgently required. *Table 2.2a* gives the equivalent values of centistokes (used in many continental countries); Redwood No. 1 Seconds (UK); Saybolt Universal Seconds (USA); and Degrees Engler (Germany). From an applications point of view centipoises (centistokes × density) are probably more useful and in the USA the reyn (1 centipoise =

Table 2.2a VISCOSITY CONVERSION, CENTISTOKES, REDWOOD NO. 1, SAYBOLT UNIVERSAL, ENGLER. {FOR REDWOOD NO. 2, REYNS AND SAYBOLT FUROL, SEE FOOTNOTES TO TABLE.}

Centi-stokes	*Redwood No. 1 secs. at*			*Saybolt univ. secs. at*			*Engler°*
	21 °C	60 °C	94 °C	37.8 °C	54 °C	98.9 °C	
4	35.3	35.9	36.3	39.1	39.2	39.4	1.32
5	37.9	38.5	38.9	42.3	42.4	42.6	1.40
6	40.5	41.1	41.5	45.5	45.6	45.8	1.48
7	43.2	43.7	44.1	48.7	48.8	49.0	1.57
8	46.0	46.3	46.9	52.0	52.1	52.4	1.65
9	48.8	49.1	49.7	55.4	55.5	55.8	1.74
10	51.7	52.0	52.6	58.8	58.9	59.2	1.84
12	57.9	58.1	58.7	65.9	66.0	66.4	2.03
14	64.4	64.6	65.2	73.4	73.5	73.9	2.23
16	71.0	71.4	72.2	81.1	81.3	81.7	2.44
18	77.9	78.5	79.3	89.2	89.4	89.8	2.66
20	85.0	85.8	86.9	97.5	97.7	98.2	2.88
22	92.4	93.3	94.5	106.0	106.2	106.7	3.10
24	99.9	100.9	102.2	114.6	114.8	115.4	3.32
26	107.5	108.6	110.0	123.3	123.5	124.2	3.57
28	115.3	116.5	118.0	132.1	132.4	133.0	3.82
30	123.1	124.4	126.0	140.9	141.2	141.9	4.08
32	131.0	132.3	134.1	149.7	150.0	150.7	4.34
34	138.9	140.2	142.2	158.7	159.0	159.8	4.59
36	146.9	148.2	150.3	167.7	168.0	168.9	4.85
38	155.0	156.2	158.3	176.7	177.0	177.9	5.10
40	163.0	164.3	166.7	185.7	186.1	187.0	5.36
45	183.1	184.5	187.5	208.4	208.8	209.9	6.00
50	203.3	204.7	208.3	231.4	231.8	233.0	6.58
60	243.5	245.3	250.0	277.4	277.9	279.3	7.89
70	283.9	286.0	291.7	323.4	324.0	325.7	9.21
80	323.9	326.6	333.4	369.6	370.3	372.2	10.5
90	364.4	367.4	375.0	415.8	416.6	418.7	11.8
100	404.9	408.2	416.7	462.0	462.9	465.2	13.2
	—	For viscosities above 100 centistokes multiply by					
>100	4.049	4.082	4.167	4.620	4.629	4.652	0.132

Notes. 1. The Redwood No. 2 scale is occasionally used to define the viscosity of very thick oils. A close approximation to the equivalent Redwood No. 1 figure may be obtained by multiplying the Redwood No. 2 figure by 10.

2. Similarly there is an alternative Saybolt scale known as *Saybolt Furol* where the approximate Saybolt Universal figure will be obtained by multiplying the Furol figure by 10.

3. For absolute viscosity in centipoise multiply centistokes by the density of the fluid. To convert centipoise to reyns ($lb \cdot sec/in^2$) multiply centipoise by $1.45 \cdot 10^{-7}$.

$1.45 \cdot 10^{-7}$ reyns) is often preferred since its units are lb(force) × sec/in^2.

To add to the general confusion the automobile industry uses a rather approximate system of SAE numbers and the equivalent of these in Redwood No. 1 seconds at 60 °C are shown in *Table 2.2b*.

Table 2.2b S.A.E. NUMBER EQUIVALENTS IN REDWOOD NO. 1 SECONDS AT 60 °C

SAE No.	*Redwood No. 1 Seconds* (60 °C)
10 (extra light)	70–90
20 (light)	90–130
30 (medium)	130–175
40 (medium heavy)	175–250
50 (heavy)	250–375
60 (extra heavy)	375–500
70 (super heavy)	Greater than 500

Polymeric additives are sometimes used to improve the viscosity/temperature characteristics of the oil and normally extend the operating temperature range. Oils having a viscosity index (see 20 on p. 27) greater than about 120 usually contain these additives. When such oils are subjected to high shear rates *reversible* and *irreversible* reductions in viscosity and viscosity index may occur. The amount of the reduction will depend upon the type of additive used, its molecular weight, the rate of shear applied to the fluid, the period of time during which it is subjected to the shear rate and the oil temperature.[7] With all these variables it is naturally impossible to generalise, but reductions in viscosity of between 35 and 45% have occurred to a service type fluid after 10 000 circulations in a pump system at pressures between 140 and 230 bars.[8] Tests for permanent shear breakdown using a pump are time consuming and alternative accelerated tests have been suggested, but no standard test method has been established. A method proposed by Wood[9] is now widely used in which the fluid is continuously passed through a diesel injector nozzle. An alternative proposed method in which the fluid is subjected to irradiation in a sonic oscillator is described by the ASTM.[3]

18. *Pressure and temperature effects.* The viscosity μ of oils may be described by the equation

$$\mu = \mu_o e^{\alpha P - \beta T}$$

where μ_o is the viscosity at some datum pressure and temperature and α and β are pressure and temperature coefficients and P and T are the oil pressure and temperature measured from their datum values. The temperature coefficient is generally only constant for changes of temperature not exceeding approximately 25 °C although this depends to some extent on the type of fluid considered. Typical values of α and β are $2.2 \cdot 10^{-3}$ bars^{-1} and $3.5 \cdot 10^{-2}$/°C (Shell Tellus 27 oil) and both Bondi[10a] and Fuller[10b] give further details

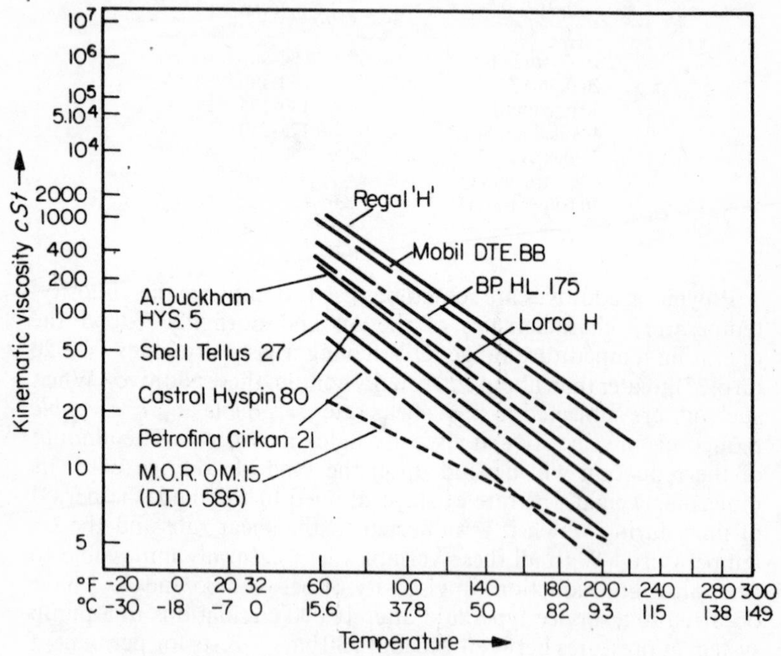

Figure 2.3. Variation of kinematic viscosity with temperature

of these quantities whilst Worster[10c] has analysed the results presented in more than fifty papers on this subject.

Examples of the very rapid reduction in the viscosity of most mineral-based hydraulic fluids with increasing temperature are illustrated in *Figure 2.3*, which shows the results for a selection of oils. The scales of the ASTM chart on which these results are presented are specially selected and the vertical scale is a log-log one in order to cover the very wide range which is sometimes required. The smaller but nevertheless important effect of pressure on viscosity is illustrated in *Figure 2.4* where it can be seen that the viscosity of an oil is approximately doubled when the pressure rises from 0 to 300 bars.

19. *Shear breakdown.* Both the temporary and permanent reductions which occur in the value of the viscosity of a fluid are associated with the long chain molecules of various additives. The

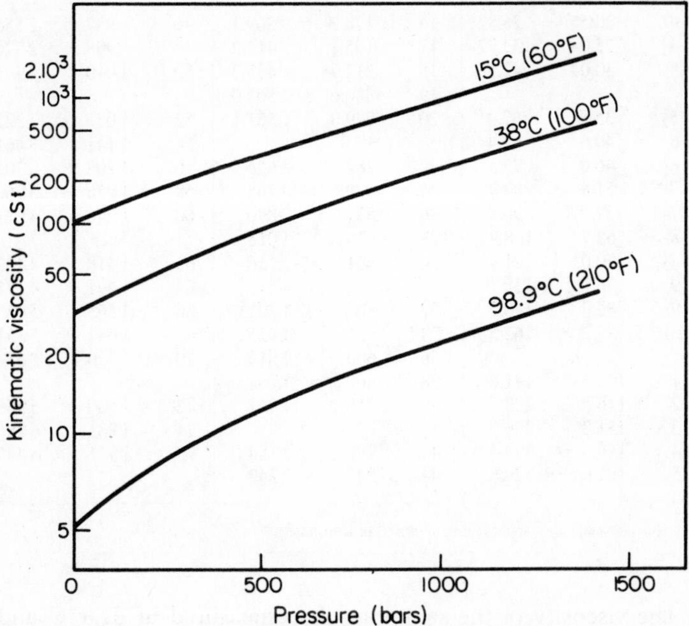

Figure 2.4. Variation of the kinematic viscosity with pressure

temporary reduction is thought to be caused by the 'lining-up' of the long molecules in a plane parallel to the direction of shear whilst the permanent reduction is thought to be due to the mechanical breaking of these long chains such that their effect of increasing the viscosity is permanently lost.

20. *Viscosity Index* (VI). Unfortunately, although the idea behind the use of the viscosity index (VI), as a measure of the change of an

oil's viscosity with temperature is sound, it is not a simple test to use since in addition to the determination of an oil's viscosity at 37.8 °C and 98.9 °C (100 °F and 210 °F) it also requires the use of tables giving reference quantities. Initially, an oil's viscosity, v, in centistokes is determined at 98.9 °C and, from tables, oils having VI values of 0 and 100 are found that have the same value of viscosity as the test sample at 98.9 °C.

Table 2.3 DATA FOR ESTIMATING VISCOSITY INDEX

v	H	L	v	H	L	v	H	L
4	20.55	26.82	16	178.2	369.1	46	887	2 555
4½	25.04	33.72	17	195.1	411.2	48	948	2 770
5	30.04	42.56	18	212.7	455.7	50	1 010	2 991
			19	230.6	502.0			
5½	35.3	52.4	20	249.3	551.1	52	1 075	3 222
6	40.6	62.4				54	1 141	3 461
6½	46.0	72.5	22	288	655	56	1 207	3 705
7	51.8	83.9	24	329	768	58	1 276	3 960
7½	57.7	96.5	26	371	890	60	1 346	4 221
8	63.7	108.9	28	415	1 018			
8½	70.0	121.9	30	461	1 156	62	1 418	4 492
9	76.3	135.2				64	1 491	4 771
9½	82.7	148.7	32	508	1 301	66	1 565	5 055
10	89.2	162.5	34	557	1 455	68	1 642	5 351
			36	609	1 619	70	1 720	5 654
11	102.5	191.8	38	660	1 788			
12	116.5	223.1	40	715	1 967	72	1 800	5 966
13	131.2	256.7				74	1 881	6 286
14	146.4	292.4	42	770	2 154	75	1 921	6 447
15	161.9	329.6	44	827	2 349			

(v = kinematic viscosity of sample at 98.9 °C in centistokes).

The viscosity of the sample is then measured at 37.8 °C and if its value is U and the viscosities of the two reference oils at 37.8 °C are L for the oil of zero VI and H for the oil of 100 VI then the viscosity index of the sample oil is given by

$$\mathrm{VI} = \left\{\frac{L-U}{L-H}\right\} \times 100$$

This is shown diagrammatically in *Figure 2.5* and *Table 2.3* gives a condensed version of the information in IP 73/42 and may be used to obtain values of *L and H* (by interpolation if necessary) provided the cloud point of the fluid under test is below 37.8 °C.

A high VI value is generally desirable since it denotes an oil the viscosity of which does not change greatly with temperature. Most

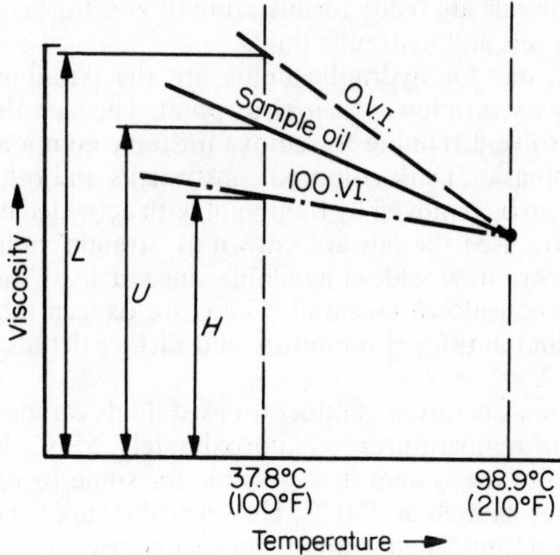

Figure 2.5. Determination of viscosity index

modern mineral based oils have values of between 80 and 95 but additives can be used such that values of up to 150 can be achieved.

2.3 Mineral-based hydraulic fluids

This type of fluid is the most commonly used in the majority of hydraulic systems and is, therefore, conveniently used as a basis for comparison when considering the behaviour of all other types of fluid.

2.3.1 Production and characteristics

Mineral-based oils are compounded from selected lubricating oils and small amounts of chemical additives. The latter are usually chosen to enhance particular aspects of various properties of the base oil, which is the oil chosen to form the major part of the fluid. The crude oils from which the base oils come vary widely in composition and usually consist of mixtures of one of three main groups of hydrocarbons; aromatics, naphthenes and paraffins. The last two groups make the best lubricating oils.

Fractional distillation is used to refine the crude oil and further refinement of some of the heavy residues left after this process produces unrefined lubricating oils. After a further refining process the resulting oils are ready for blending to give the general range of lubricating oils and hydraulic fluids.

The best oils for hydraulic fluids are the paraffins, which are usually dewaxed to lower their pour point. They are then subjected to acid or solvent refining to remove unstable components such as certain aromatics. Dark coloured constituents and other unwanted materials can be removed by treatment with activated earths. Before additives are used the oils are known as 'straight' mineral oils but such fluids are now seldom available since at least three additives are usually considered essential. These are oxygen inhibitors, rust inhibitors and anti-foam inhibitors and further details are given in section 2.3.2.

Satisfactory operation of mineral-based fluids can be obtained up to reservoir temperatures of approximately 55 °C, but if air is excluded from the system it is possible for some to operate up to temperatures as high as 150 °C. However they are very susceptible to oxidation and their vapour pressure rises rapidly at high temperatures so that rapid deterioration, evaporation and the resulting pump cavitation usually prohibits their use at high temperatures. When the search for fluids for high temperature applications began it was found that it was possible to operate mineral based fluids at high temperatures after further refining using silica gel absorption or hydrogenation over a nickel catalyst.[11] These processes greatly assisted the action of the oxidation and corrosion inhibitors above 180 °C, and although working temperatures up to 280 °C became feasible the fire resistant or non-flammable fluids are nowadays preferred.

Special sealing materials such as nitrile, Neoprene and polysulphides have been developed for use with mineral-based fluids, which must not be used with natural rubber seals (see *Table 4.2*. page 75). The aniline point plays an important part in determining the behaviour of a fluid in contact with a sealing material and it can also be used to identify the crude oil from which the fluid originally came.

Standard oils are used as a basis of comparison for the effect of oils on synthetic rubbers. The first (No. 1) oil has an aniline point of 107 °C and causes hardening and shrinkage of one type of nitrile rubber whereas the No. 3 oil, with an aniline point of 70 °C causes it to swell and soften. It does not follow that all rubbers will respond to this test in this manner and even nitrile rubbers are classified by the terms 'high swell', 'medium swell' and 'low swell' which describes their behaviour under this test. Neoprene generally swells when in contact with mineral oils and polysulphides are not greatly affected by volume changes but they do have other practical limitations.

2.3.2 Additives

These substances are generally added in very small quantities to the base fluid either to enhance or inhibit certain of its characteristics. Additives that are commonly used are:

(a) *Anti-oxidant.* Oxidation inhibitors may be added for high temperature and high pressure use and these delay the onset of oxidation. The quantity used is generally between 1 and 2% by volume.

(b) *Rust inhibitor.* This is added to enable the fluid to coat ferrous metals with a strongly absorbed oily film thereby preventing the access of moisture and air.

(c) *Anti-foam inhibitor.* Silicones and similar additives may be used to cause rapid dispersal of foam by providing a weak link of molecular thickness to induce bubble film breakage.

(d) *Extreme pressure.* These additives, sometimes referred to as EP additives, are not generally used or desirable in hydraulic fluids but are used with gear oils where very heavy loads and high temperatures are encountered. Substances which have proved to be effective in assisting the formation of a film of low shear strength on metals are chlorine, sulphur and phosphorus. Compounds containing these elements are used to react with the metal surfaces to form chlorides, sulphides and phosphides. Fairly high temperatures are needed for the reaction to take place, and the rate of reaction depends on the application. An additive which is too active under operating conditions may well do more harm than good if the high reactivity of the additives results in excessive corrosion. Additives containing sulphur may have a severe effect on standard seals, particularly at high temperatures. Sulphur-containing EP additives are alkyl polysulphide, sulphurised terpene, benzyl disulphide and sulpho-chlorinated olefin.

(e) *Viscosity index improvers.* Acryloid polymers may be added to give a more favourable viscosity-temperature characteristic to low viscosity fluids.

(f) *Anti-wear.* Approximately 0.5% by weight of tricresyl phosphate may be used to reduce friction and the consequent wear. It may give rise to a polishing effect which increases the load carrying area. Other lubricity additives are polybutene and dialkyl hydrogen phosphate and trixylyl phosphate.

Table 2.4 SOME DETAILS OF COMMERCIALLY AVAILABLE MINERAL OILS

Manufacturer and Oil	*Viscosity, cSt (IP 71/62)*			*Viscosity Index (IP 73/53)*	*Specific Gravity (IP 59/57)*		*Aniline Point °C*	*Additives**
	−12.2°C	*37.8°C*	*98.9°C*		*15.6°C*	*98.9°C*		
B.P. Energol								
HL 40	—	9.92	2.38	50	0.876	—	87	1,2,3,6.
HL 65	—	32.7	5.18	95	0.875	—	96	1,2,3,6.
HL 80	—	44.9	6.28	95	0.878	—	97	1,2,3,6.
HL 100	—	61.6	7.65	95	0.881	—	103	1,2,3,6.
HL 150	—	101.7	10.62	95	0.882	—	110	1,2,3,6.
HL 175	—	120	12.24	100	0.888	—	110	1,2,3,6.
Castrol Hyspin								
Hyspin 40	—	10.2	2.5	71	0.863	0.811	87	1,2,3.
Hyspin 55	—	23.5	4.1	74	0.874	0.824	99	1,2,3.
Hyspin 70	—	34.0	5.4	95	0.875	0.825	98	1,2,3.
Hyspin 80	—	46.0	6.3	92	0.873	0.823	105	1,2,3.
Hyspin 100	—	61.4	7.7	97	0.882	0.832	104	1,2,3.
Hyspin 140	—	95.0	10.2	96	0.881	0.831	112	1,2,3.
Hyspin 175	—	120	11.8	94	0.887	0.838	112	1,2,3.
A. Duckham								
Hys. 1	—	6	—	—	0.830	—	—	1,2.
Hys. 2	—	22	4.2	104	0.865	—	—	1,2.
Hys. 3	—	32	5.2	102	0.875	—	—	1,2.
Hys. 4	—	52	6.9	96	0.885	—	—	1,2.
Hys. 5	—	72	8.4	94	0.890	—	—	1,2.
Hys. 6	—	112	11.5	98	0.890	—	—	1,2.
Hys. 7	—	122	12	95	0.890	—	—	1,2.
Hyq. 2	—	22	4.6	143	0.865	—	—	1,2,5.
Hyq. 3	—	32	6.2	146	0.865	—	—	1,2,5.
Hyq. 4	—	47	8.2	140	0.865	—	—	1,2,5.
Hyq. 5	—	62	10	137	0.870	—	—	1,2,5.
Hyq. 6	—	82	12	132	0.875	—	—	1,2,5.

Table 2.4 *(continued)*

Manufacturer and Oil	*Viscosity, cSt (IP 71/62)* −12.2 °C	37.8 °C	98.9 °C	*Viscosity Index (IP 73/53)*	*Spedific Gravity (IP 59/57)* 15.6 °C	98.9 °C	*Aniline Point* °C	*Additives**
Lorco								
L	770	32.0	5.23	95 (min)	0.880	—	100	1,2,3,4,6.
M	1 700	47.0	6.4		0.885	—	102	1,2,3,4,6.
MH	2 600	67.5	8.4		0.885	—	103	1,2,3,4,6.
H	4 400	86	9.5		0.885	—	105	1,2,3,4,6.
XH	8 000	112	11.2		0.887	—	106	1,2,3,4,6.
XX	13 000	170	14.8		0.889	—	109	1,2,3,4,6.
Y	19 800	230	17.8		0.890	—	112	1,2,3,4,6.
YY	43 000	330	22.5		0.895	—	116	1,2,3,4,6.
A	228	13.6	2.9	70 (min)	0.875	—	73	1,2,3,4.
B	545	22.8	3.96		0.887	—	86	1,2,3,4.
L	840	33.2	5.0		0.887	—	91	1,2,3,4.
M	1 620	47.2	6.2		0.893	—	92	1,2,3,4.
MH	4 500	71	7.8		0.895	—	93	1,2,3,4.
H	7 800	97	9.2		0.898	—	93	1,2,3,4.
XH	15 000	143	11.7		0.902	—	94	1,2,3,4.
XX	49 000	249	15.5		0.900	—	97	1,2,3,4.
M.O.R.								
46 IH	260	13.6	2.97	69	0.875	0.821	87	1,2,3.
68 IH	1 500	35.8	5.11	72	0.888	0.834	90	1,2,3.
80 IH	2 800	50	6.26	73	0.892	0.838	92	1,2,3.
100 IH	5 300	69	7.60	98	0.895	0.841	95	1,2,3.
OM 13	260	13.6	2.97	69	0.875	0.821	87	
OM 15	75	14.3	5.1		0.862	0.808	77	

Table 2.4 *(continued)*

Manufacturer and Oil	*Viscosity, cSt (IP 71/62)*			*Viscosity Index (IP 73/53)*	*Specific Gravity (IP 59/57)*		*Aniline Point °C*	*Additives**
	−12.2 °C	*37.8 °C*	*98.9 °C*		*15.6 °C*	*98.9 °C*		
Mobil								
D.T.E. Light	—	32	5.26	103	0.870	—	—	1,2,3.
D.T.E. Medium	—	48	6.92	100	0.875	—	—	1,2,3.
D.T.E. Medium Heavy	—	63	8.10	97	0.880	—	—	1,2,3.
D.T.E. Heavy	—	90	10.19	95	0.890	—	—	1,2,3.
D.T.E. Extra Heavy	—	130	12.93	95	0.890	—	—	1,2,3.
D.T.E. BB	—	197	16.52	95	0.900	—	—	1,2,3.
D.T.E. AA	—	388	26.0	95	0.900	—	—	1,2,3.
D.T.E. LC	—	35.0	7.4	150	0.867	—	—	1,2,3,5,6.
Vacuo-Line 1405	—	32.0	5.1	100	0.875	—	—	4.
Vacuo-Line 1409	—	65	7.3	75	0.895	—	—	4.
Petrofina								
Cirkan 11	—	4.62	—	—	0.836	—	—	1,2,3.
Cirkan 15	—	9.77	2.38	56	0.866	—	—	1,2,3.
Cirkan 21	—	23.05	4.00	63	0.881	—	—	1,2,3.
Cirkan 31	—	35.61	5.22	71	0.892	—	—	1,2,3.
Cirkan 32	—	45.00	5.90	75	0.892	—	—	1,2,3.
Cirkan 37	—	70.90	8.10	88	0.888	—	—	1,2,3.
Cirkan 43	—	94.00	9.90	91	0.880	—	—	1,2,3.
Cirkan 46	—	109.90	11.20	95	0.885	—	—	1,2,3.

Table 2.4 *(continued)*

Manufacturer and Oil	*Viscosity, cSt (IP 71/62)*			*Viscosity Index (IP 73/53)*	*Specific Gravity (IP 59/57)*		*Aniline Point °C*	*Additives**
	−12.2 °C	*37.8 °C*	*98.9 °C*		*15.6 °C*	*98.9 °C*		
Petrofina								
Hydran 11	—	6.32	1.97	110	0.825	—	—	1,2,3.
Hydran 12	—	10.80	2.75	107	0.843	—	—	1,2,3.
Hydran 21	—	18.00	3.74	107	0.857	—	—	1,2,3.
Hydran 31	—	34.30	5.50	107	0.866	—	—	1,2,3.
Hydran 32	—	44.80	6.56	107	0.869	—	—	1,2,3.
Hydran 34	—	54.00	7.40	106	0.870	—	—	1,2,3.
Hydran 37	—	65.00	8.34	106	0.875	—	—	1,2,3.
Hydran 43	—	82.00	9.60	105	0.875	—	—	1,2,3.
Hydran 46	—	115.50	11.79	102	0.875	—	—	1,2,3.
Hydran 51	—	132.37	13.12	100/101	0.880	—	—	1,2,3.
Hydran 53	—	184.43	16.48	101/102	0.884	—	—	1,2,3.
Hydran 58	—	224.36	17.79	93/94	0.887	—	—	1,2,3.
Petrofina								
Hydran (125) 11	—	6.44	2.08	132/133	0.830	—	—	1,2,3,5.
Hydran (125) 12	—	9.21	2.60	130	0.838	—	—	1,2,3,5.
Hydran (125) 21	—	22.24	4.22	124/125	0.838	—	—	1,2,3,5.
Hydran (125) 31	—	38.43	6.55	130/131	0.872	—	—	1,2,3,5.
Hydran (125) 32	—	42.75	6.97	127	0.875	—	—	1,2,3,5.
Hydran (125) 34	—	47.37	7.53	126/127	0.876	—	—	1,2,3,5.
Hydran (125) 37	—	57.43	8.81	129/130	0.876	—	—	1,2,3,5.
Hydran (125) 43	—	77.30	10.82	126/127	0.876	—	—	1,2,3,5.
Hydran (125) 46	—	99.30	13.32	128	0.877	—	—	1,2,3,5.
Hydran (125) 51	—	113.10	17.20	129/130	0.878	—	—	1,2,3,5.
Hydran (125) 53	—	165.20	20.83	130/131	0.879	—	—	1,2,3,5.
Hydran (125) 58	—	195.30	23.27	126/127	0.881	—	—	1,2,3,5.

Table 2.4 *(continued)*

Manufacturer and Oil	*Viscosity, cSt (IP 71/62)* −12.2 °C	37.8 °C	98.9 °C	*Viscosity Index (IP 73/53)*	*Specific Gravity (IP 59/57)* 15.6 °C	98.9 °C	*Aniline Point* °C	*Additives**
Regent								
Regal A	—	32.7	5.2	97	0.867	—	101	1,2,3.
Regal B	—	45.7	6.37	95	0.873	—	105	1,2,3.
Regal PC	—	68	8.1	94	0.877	—	113	1,2,3.
Regal PE	—	90.6	9.62	91	0.883	—	114	1,2,3.
Regal F	—	145.6	12.83	86	0.889	—	112	1,2,3.
Regal G	—	211.6	17.93	100	0.884	—	121	1,2,3.
Regal H	—	276.6	20.35	94	0.887	—	123	1,2,3.
Shell								
Tellus 11	31	6.0	—	95/100	0.830	—	88	1,2,3,4.
Tellus 15	200	12.7	—	95/100	0.855	—	97	1,2,3,4.
Tellus 27	870	34.5	5.6	95/100	0.975	—	108	1,2,3,4.
Tellus 29	1 800	49.5	7.0	95/100	0.875	—	107	1,2,3,4.
Tellus 33	3 800	70.0	8.6	95/100	0.880	—	111	1,2,3,4.
Tellus 41	6 000	108	11.8	95/100	0.885	—	113	1,2,3,4.
Tellus 69	—	141	14.0	95/100	0.888	—	116	1,2,3,4.
Tellus 72	—	208	18.2	95/100	0.890	—	125	1,2,3,4.
Tellus T17	180	15.7	3.8	—	0.869	—	90	1,2,3,4,5.
Tellus T27	750	45.2	8.7	132	0.873	—	95	1,2,3,4,5.
Aeroshell 4 (D.T.D. 585)				200				
Edgar Vaughan								
Evco Light	210	14.7	3.28	100	0.875	0.827	87	1,2.
Evco Medium	970	35.1	5.54	104	0.879	0.829	101	1,2.
Evco Heavy	2 100	53.0	7.1	100	0.879	0.829	103	1,2.
Evco Extra Heavy	—	66.0	8.2	100	0.881	0.830	105	1,2.

* Types of additives present.
1. Anti-oxidant,
2. Anti-foam,
3. Anti-rust
4. Anti-wear,
5. V.I. Improver,
6. Pour-point depressant

(g) *Pour point depressants.* These may be added to reduce the onset of solidification at lower temperatures and are often necessary with high viscosity index oils.

(h) *Anti-deterioration.* Chemicals may be added to synthetic fluids to increase their storage life to three years.

(i) *Gum solvency or detergent.* These can help to maintain a clean system by dissolving precipitates. A typical detergent is 'Tergol' and other detergents are derived from alkylated salicylic acid.

(j) *Multifunctional.* These additives may perform several of the actions described above at the same time. Details of several commercially available mineral oils are given in *Table 2.4*.

2.4 Castor-based hydraulic fluids

Also known as vegetable-based oils by reason of their origin the use of these fluids has rapidly declined in recent years. They were developed for use with natural rubber sealing materials which are particularly resilient and flexible at low temperatures and are strong and resistant to abrasion.

2.5 Synthetic hydraulic fluids

The silicate esters, polysiloxanes and disiloxanes are used for systems having a large operating temperature range although their flammability characteristics are still open to question.[12] Their greatest limitation is their susceptibility to water contamination which produces complete hydrolisis at a 0.5% concentration at temperatures above 205 °C. They perform well with standard synthetic sealing materials since their tendency to shrink can be countered by blending with diesters but they do have a tendency to corrode some metals commonly used in system construction at high temperatures.

2.6 Non-flammable fluids

The increasing demand for systems to operate at high temperatures has led to the development of several special types of fluids which are now usually classified under the general heading of 'non-flammable'.

The following four main types are now generally recognised:

(a) Water based; water glycols; water in oil emulsions.
(b) Phosphate ester based.
(c) Silicones.
(d) Halogenated hydrocarbons (chlorinated and fluorinated hydrocarbons).

Fluids in the first group are fire resistant by virtue of the presence of the water, which acts as a snuffing agent and fluids in the other groups are chemically resistant to burning or oxidation. However, there are very few materials which cannot be made to burn when given the right conditions, and for this reason the following additional factors are usually considered:

(a) Will a flame flash back to the point at which the system ruptured?
(b) Will a spray or mist of the fluid explode in the presence of a flame?
(c) Will a flame be self supporting away from the immediate source of heat?

The results of most tests show that water based fluids will not generally burn until the water content has evaporated and that synthetic fluids will only burn in the region of a point maintained at or above the fluid fire point, the combustion being entirely localised.

The *standard* fire and flash point tests used with general hydraulic fluids *have little meaning* when applied to non-flammable fluids and other tests approximately simulating practical conditions are preferred. These are the spraying of a fluid jet through an oxy-acetylene torch or over a burning rag, the repeated passing of a saturated pipe cleaner through a flame and the dripping of the fluid onto a hot metal block (the manifold test). Other tests which are also used are described in Reference 13.

2.6.1 Water glycols

The percentage of water present in these fluids varies between 30 and 60% and the remainder of the fluid is ethylene glycol with a little thickening agent to give the required viscosity. Lubricity, anti-wear and anti-corrosion additives must usually be added but even with these present water glycols should not be used with magnesium, zinc or cadmium. It should also be remembered that they will soften most types of paint. Although such fluids can be used in vane pumps and also also in piston and gear pumps if peak loads are avoided it must be noted that their lubricating properties are usually very inferior to those of mineral and castor-based oils. Maximum pressures of 70 bars for vane pumps and 30 bars for other types of pump are generally accepted.

Water glycols have a satisfactory, low temperature viscosity and a viscosity index of approximately 150, but temperatures above 55 °C must be avoided to ensure that the water content and volatile inhibitors do not evaporate. Special water[14] may be used to make up any evaporation losses and chemical filters should be avoided since they will usually remove the anti-corrosion additives. Standard mineral oil seals may be used with such fluids but contamination with certain solvents such as *carbon tetrachloride must be avoided.*

2.6.2 Water-oil emulsions

These fluids consist of a low viscosity mineral oil in which water is dispersed in fine droplets. They are generally not so reliable as water glycols, since they can become unstable, although their performance in pumps is similar and they may be used at somewhat higher pressures. Their comparatively low cost makes them attractive to the mining industry where leakage losses can sometimes be excessive.

Their operating life can be measured in years if the operating temperature can be kept as low as 25 °C but it decreases rapidly with rising temperature which must not exceed 55 °C.

Contact with any one of several other types of fluid can result in the emulsion becoming unstable and such fluids include solvents, chemical cleaners, and synthetic hydraulic fluids such as phosphate esters. In addition a small trace of water glycol is sufficient to separate the oil from the water. Any restriction to flow at low pressures (suction lines and low pressure filters) is liable to cause separation. If separation has occurred after a long shut down period, remixing is generally a simple matter of providing sufficient agitation.

In common with water glycols the relatively high specific heat and thermal conductivity of such fluids reduces overheating problems. They are non-toxic, non-corrosive, have no effect on paints and may be used with mineral oil seals provided cork and high sulphur content compounds are absent.

2.6.3 Phosphate ester based fluids

A chemical reaction between phosphoric acid and an alcohol[15] produces phosphate esters, which are sometimes mixed with chlorinated hydrocarbons and are probably the most widely used synthetic fluids. They will only burn at the point of ignition when the fluid is maintained above its fire point and the flame will not flash back to the point of rupture. This is because the fluid must first be decomposed into combustile products before burning can commence.

Pump manufacturers' test results suggest that phosphate esters may be used in vane pumps up to 100 bars and in piston pumps up to 130 bars. They also indicate that the fluids have satisfactory lubricity and wear characteristics, but their higher specific gravity can aggravate suction line problems. Provided an inhibitor additive is used, they do not corrode system materials but they do attack paints and some types of electrical insulation materials.

In spite of a high initial cost they are said to be economical since they have a long service life and may be reclaimed by filtration. Their low viscosity index should be noted if use over a wide temperature range is envisaged but their poor, low temperature characteristics are often outweighed by their excellent, high temperature ones. Maximum, operating temperatures up to 150 °F are quoted and up to this point phosphate esters have high thermal and oxidation stability.

Although their hydrolytic stability is good since they are both non-soluble and non-miscible, it is essential to avoid water contamination. At high temperatures this can produce an acidic reaction with copper and its alloys.

All phosphate ester fluids should be treated carefully and handled with caution. Those containing chlorinated hydrocarbons emit chlorine fumes and others are toxic in the vapour state. As a result adequate ventilation should always be provided if such fluids are open to the atmosphere. In addition they have strong solvent properties and can cause defatting of the skin, so that rubber gloves or at least barrier cream must be used when working on systems containing these fluids.

Standard nitrile seals should not be used with phosphate esters and it is generally advisable to consult a seal manufacturer before selecting sealing materials. Generally silicone, nylon, leather, ptfe and butyl are satisfactory but the latter swells if a chlorinated hydrocarbon is present. In addition, since butyl is incompatible with mineral oil there must be no trace of the latter in phosphate esters. As a general rule it is essential to change all seals if a phosphate ester is replaced by a mineral oil or vice-versa.

'Cellulube'[16a] is a trade name given to pure phosphate esters and since no chlorinated hydrocarbons are added they do not emit chlorine fumes. In addition, they contain no thickeners and as they require no additives maximum system filtration[16b] may be used without changing the fluid properties.

An outstanding point in favour of phosphate esters is their high bulk modulus which can be between $1\frac{1}{2}$ and $3\frac{1}{2}$ times that of mineral based fluids.

2.6.4 Silicones

The fire resistance of silicones covers a wide range and their lubricity particularly for steel on steel is poor. However, new base materials are being developed with a view to improving their lubricating properties since their great attraction is a very high thermal stability and viscosity index. The use of additives provides little solution to the problem for generally only small improvements are obtained with the few which are miscible with silicones and even these do not usually enjoy the same resistance to high temperatures.

'Sulfiniz' treatment of the pistons of a radial piston pump enabled a machine to be operated with a silicone fluid at pressures as high as 300 bars with satisfactory wear rates[17a] and this is also a field in which development is taking place. The results of the life testing of pumps and motors using various water based fluids are described by Platt and Kelly.[17b]

Sealing difficulties are often encountered since silicones usually have a low surface tension and shrink conventional rubbers. This shrinkage can be overcome by the use of certain additives.

Their low bulk modulus is yet another unattractive feature, but it is almost certain that each of the undesirable properties mentioned above will eventually be eliminated and the reader should endeavour to keep abreast with recent developments in this field.

2.6.5 Chlorinated hydrocarbons

Belonging to the halogenated hydrocarbon family and including fluids known as chlorinated diphenyls, these fluids are generally only used as heat transfer liquids[15] since they are expensive. Some are virtually non-flammable and all are very fire-resistant, having a maximum operating temperature of about 260 °C. Although they are available in a wide range of viscosities they have a low viscosity index and a very high density.

Their lubricity and wear properties are usually so excellent that they are sometimes used as extreme pressure (EP) additives for lubricating oils, in addition to their use with phosphate esters mentioned earlier. However, Hatton[18] does have certain reservations concerning their universal applications. With the exception of their effect on copper at high temperatures they are non-corrosive with hydraulic system materials and have good thermal and hydrolitic stability, although they attack paint.

At room temperature they are virtually non-toxic but care should be taken in handling them since they can be absorbed through the skin and in the vapour form become highly toxic.

Only fluorinated rubbers, silicones, nylon and ptfe, seals may be used with these fluids.

2.6.6 Fluorinated hydrocarbons

Being almost identical in properties to chlorinated hydrocarbons, these fluids are not often used because of their high cost. If aluminium threaded connections are present they should not be used since rubbing contact with this metal can initiate reaction.

REFERENCES

1. '*I.P. Standards for petroleum and its products*'. Institute of Petroleum (March, 1960).
2. Perks, B. and Westcott, M. J., *Pump tests at high temperature on a synthetic hydraulic fluid with colloidal additives.* R.A.E. Tech. Note; ME, 254. (February, 1958).
3. 'A.S.T.M. Standards on petroleum products and lubricants'. Am. Soc. Test. Mat. **1.** (October, 1961).
4. Shute, N. A. 'Data on Mineral-base Hydraulic Fluids'. B.H.R.A., T.N. 756. (March, 1963).
5. Hayward, A. T. J., 'Aeration in hydraulic systems—its assessment and control'. *I.Mech.E. Proc. Conf. Oil Hydraulic Power Transmission*, p. 216–24. (November, 1961).
6. Khoklov, B. A., 'Electrohydraulically Controlled Drives', 1st Ed., Hayka, Moscow (1964).
7. Porter, R. S. and Johnson, J. F., 'Flow properties of modern lubricating oils'. *J.A.S.N.E.* **73,** 3, p. 511–5. (August, 1961).
8. Vaughan, I. J. and Westcott, M. J., 'The effect of pressure on the shear stability of mineral-base hydraulic fluids'. R.A.E., Mem, M.E. 65. (March, 1955).
9. Wood, L. G., 'The change of viscosity of oils containing high polymers when subjected to a high rate of shear'. B.J. Ap. Phy. p. 206–8. (August, 1950).
10. Bondi, A., *Physical Chemistry of Lubricating Oils*, 1st edition, Reinhold, New York (1951).

10b. Fuller, D. D., *Theory and Practice of Lubrication for Engineers*, 1st edition, Wiley, New York (1956).

10c. Worster, R. C., *The Effect of Pressure on the Viscosity of Liquids and Especially Oils.* B.H.R.A., T.N.16 (1950).
11. Klaus, E. E. and Fenske, M. R., 'High temperature lubricant studies'. *Lub. Engg.* p. 266–73. (June, 1958).
12. Blackburn, J. E. et al., *Fluid Power Control.* 1st Edition. J. Wiley and Sons (1960).
13. *DTD. 5507. Aircraft Material Specification, Hydraulic Fluid—Low Inflammability.* Third Draft (April, 1956).
14. Shute, N. A., 'A review of hydraulic fluids'. *B.H.R.A.*, T.N. 673. (October, 1960).
15. 'The technology of oil sealing materials'. Part 9. 'Fire resisting hydraulic fluids'. Oilseal Jnl. (Pioneer Oilsealing and Moulding Co. Ltd.) No. 17, p. 193–5. (March, 1960).
16a. Matthews, W. D., 'Fire resistant hydraulic fluids. Pt. III. Pure phosphate esters'. *App. Hyd.* **10,** No. 9, p. 168–73. (September, 1957).
16b. Denny, D. F., 'Cleanliness in Hydraulic Systems'. *Proc. Oil Hyd. Pow. Trans and Cont., I.Mech.E.* (November, 1961).
17a. Perks, B. and Wheeler, E. J. R., 'Pump wear tests to specification D.T.D. 5507 on G.E. 8-1406 synthetic hydraulic fluids'. R.A.E. Tech Note; M.E. 263. (June, 1958).
17b. Platt, A. and Kelly, E. S., 'Life testing of hydraulic pumps and motors on fire resistant fluids'. B.H.R.A., SP 982 (January, 1969).
18. Hatton, R. E., *Introduction to Hydraulic Fluids.* 1st edition. Reinhold, New York. (1962).

References 4 and 14 are of particular value since between them they contain nearly 100 references on the subject of hydraulic fluids and note should also be taken of reference 10a which contains a large quantity of scientific data.

Chapter 3

Bearings

3.1 Introduction

In all machines and mechanisms contact between components gives rise to friction, loss of mechanical power, and wear of the components themselves. It is the function of lubrication to interpose a film of lubricant between the components to improve the efficiency and life of the working mechanism.

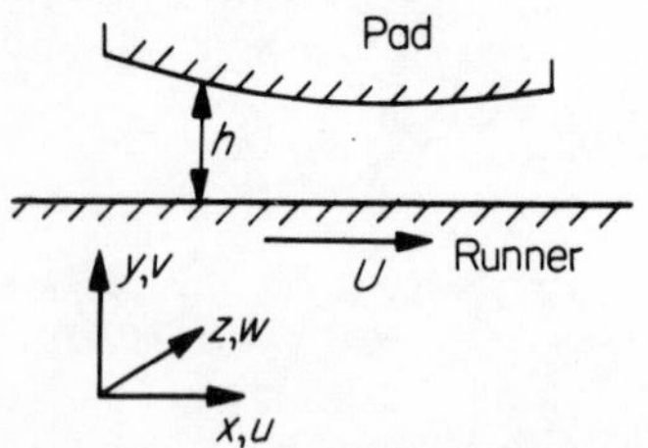

Figure 3.1. Notation for slider pad

The lubricant film may be a solid film or grease intended to survive the life of the components it lubricates, or it may be a fluid which is periodically or continuously replenished. Typical practical examples of the latter are the slipper pad and valve plate bearings for axial piston machines discussed in Chapter 5. If the components are subject to sliding speeds which give rise to high rates of heat generation it is necessary to use a system of continuous circulation of the lubricant so that heat can be removed from the source of its generation.

The advantages of fluid-film lubrication are that so long as the relatively moving components are separated by a film of lubricant, wear of the components is not possible and the frictional resistance is limited to the force required to shear the fluid film. The separation of the components requires that there be a pressure distribution within the fluid film so that the force between one component and another can be transmitted without mechanical contact between the components. The means whereby the required pressure may be provided within the lubricant film is the subject of fluid film lubrication. Two branches of the subject may, for convenience, be identified according to the means whereby pressurisation of the film is achieved.

(a) *Hydrodynamic lubrication*, in which the pressure in the film is induced by the motion of one component surface relative to another.
(b) *Hydrostatic lubrication*, in which the fluid is supplied under pressure by an external system.

3.2 Pressure distribution in a fluid film

Consider a thin fluid film which separates a pair of nearly parallel solid surfaces. Let the lower surface coincide with the plane $y = 0$, then the upper surface will be given by $y = h$, where h is dependent upon x and z and is the local separation between the two surfaces. Let the upper surface be finite in extent and fixed in position, whilst the lower surface of infinite extent moves in its own plane in the x direction with velocity U. *Figure 3.1* represents a section through the film in the direction of sliding; the dimensions through the film thickness are grossly exaggerated.

In considering the fluid flow in the film, certain assumptions and approximations may be made for most practical bearings but to demonstrate their validity requires consideration of the Navier–Stokes equation mentioned in Chapter 1 and reference to extended works on lubrication theory should be made for a fuller treatment than is here possible.

Assumptions

1. Pressure within the film is a function only of x and z, i.e. pressure variation through the thickness of the film is negligible (h is normally of the order of 0.02 mm).
2. Stresses associated with fluid acceleration are negligible compared with those due to shearing of the fluid.
3. Fluid viscosity and density are functions only of x and z, i.e. variation of these properties through the thickness of the film is negligible.

4. Steady state conditions obtain.
5. The two surfaces which bound the film are rigid and the film geometry is therefore unaffected by the film pressures generated (deflection of the bounding surface by the pressures generated in the film forms the subject of elastohydrodynamic lubrication).

Consider an element δx, δy, δz, of fluid at some point between the two surfaces. It will be subject to stresses acting in the x direction as shown in *Figure 3.2*, together with shear stresses on the xy plane

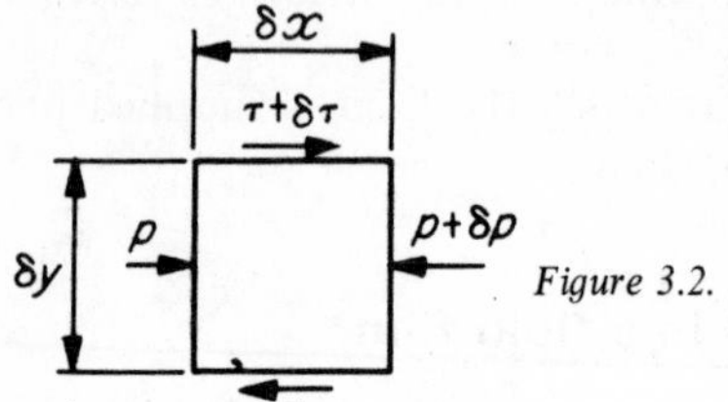

Figure 3.2. Fluid element within lubricant film

of the element which are in general very small compared with the stresses shown.

For equilibrium in the x direction:

$$p\,\delta y\,\delta z+(\tau+\delta\tau)\,\delta x\,\delta z=(p+\delta p)\,\delta y\,\delta z+\tau\,\delta x\,\delta z$$

which, in the limits as the element approaches zero size, gives

$$\frac{\partial p}{\partial x}=\frac{\partial \tau}{\partial y}$$

The shear stress τ is related to the velocity gradient $\partial u/\partial y$ by the fluid viscosity, thus

$$\tau=\mu\frac{\partial u}{\partial y}$$

since μ is considered to be independent of y

hence $$\frac{\partial p}{\partial x}=\frac{\partial}{\partial y}\left(\mu\frac{\partial u}{\partial y}\right)=\mu\frac{\partial^2 u}{\partial y^2}$$

or $$\frac{\partial^2 u}{\partial y^2}=\frac{1}{\mu}\frac{\partial p}{\partial x}$$

Similarly it may be shown that

$$\frac{\partial^2 w}{\partial y^2} = \frac{1}{\mu}\frac{\partial p}{\partial z}$$

On integrating with respect to y and applying boundary conditions

$$u = U,\ w = 0,\ \text{at } y = 0$$

$$\text{and}\quad u = 0,\ w = 0,\ \text{at } y = h$$

it follows that

$$u = \frac{1}{2\mu}\frac{\partial p}{\partial x}y(y-h) + \frac{U}{h}(h-y)$$

and

$$w = \frac{1}{2\mu}\frac{\partial p}{\partial z}y(y-h)$$

continuity of mass flow requires that

$$\frac{\partial}{\partial x}(\rho u) + \frac{\partial}{\partial y}(\rho v) + \frac{\partial}{\partial z}(\rho w) = 0$$

On substituting for u and w and integrating over the film thickness, noting that $v = 0$ at $y = 0$ and h, we obtain

$$\int_0^h \frac{\partial}{\partial y}(\rho v)\,\mathrm{d}y = 0 = -\left\{\int_0^h \frac{\partial}{\partial x}(\rho u)\,\mathrm{d}y + \int_0^h \frac{\partial}{\partial z}(\rho w)\,\mathrm{d}y\right\}$$

$$= \frac{1}{12}\frac{\partial}{\partial x}\left(\frac{h^3\rho}{\mu}\frac{\partial p}{\partial x}\right) - \frac{U}{2}\frac{\partial}{\partial x}(\rho h) + \frac{1}{12}\frac{\partial}{\partial z}\left(\frac{h^3\rho}{\eta}\frac{\partial p}{\partial z}\right)$$

$$\therefore \frac{\partial}{\partial x}\left(\frac{h^3\rho}{\mu}\frac{\partial p}{\partial x}\right) + \frac{\partial}{\partial z}\left(\frac{h^3\rho}{\mu}\frac{\partial p}{\partial z}\right) = 6U\frac{\partial}{\partial x}(\rho h) \qquad (3.1)$$

It will be evident from equation 3.1 that, within the restrictions imposed on the analysis by the assumptions made, the only self-acting pressure generation arises from the variation of density and of film thickness in the direction of sliding. Any variation of viscosity in the direction of sliding modifies but evidently does not itself give rise to pressure generation.

The solution of equation 3.1 as it stands, even for the simplest bearing geometry, is complicated by the dependence of μ and ρ on temperature. Since the temperature distribution within the film cannot be known *à priori* it is not possible to specify the variation of μ and ρ with x and z. In the case of a liquid lubricant the density variation is usually of small significance and can be ignored. The variation of viscosity however is often large due to temperature variation but, in order to render equation 3.1 tractable, viscosity will for the time being be considered constant. Reference will be made later to the consequences of this approximation. With these simplifications equation 3.1 reduces to

$$\frac{\partial}{\partial x}\left(h^3\frac{\partial p}{\partial x}\right)+\frac{\partial}{\partial z}\left(h^3\frac{\partial p}{\partial z}\right)=6\mu U\frac{\partial h}{\partial x} \tag{3.2}$$

but even in this form an analytical solution is possible for only a small number of bearing geometries and in general resort must be made to numerical integration or approximate methods of solution.

A qualitative appreciation of the performance of a bearing may be obtained analytically by neglecting the pressure gradient in the z direction, i.e. at right angles to the direction of sliding, and solving the resulting one-dimensional equation:

$$\frac{d}{dx}\left(h^3\frac{dp}{dx}\right)=6\mu U\frac{dh}{dx} \tag{3.3}$$

For a bearing in which the film thickness h is everywhere constant (except at a point of lubricant supply), equation 3.2 reduces to the Laplace equation

$$\frac{\partial^2 p}{\partial x^2}+\frac{\partial^2 p}{\partial z^2}=0 \tag{3.4}$$

and no pressure generation will result from a sliding action. Supply of lubricant under pressure will however result in a pressure distribution under the bearing whether or not there is a sliding action, and such a mode of lubrication is termed hydrostatic.

3.3 Hydrodynamically lubricated bearings

The pressure distribution in the oil film of a practical bearing or contact is two-dimensional, that is the pressure varies not only in the direction of sliding but also in the transverse direction.

Equation 3.3 above considers that there is no variation of pressure across the width of the bearing (z direction) and, since in practical bearings the sides of the bearing are exposed to the ambient pressure, this leads to over-estimation of film pressures, especially near the sides of the bearing, and therefore to over-estimation of the total load supported by the film. For the prediction of load capacity of a bearing the effect of the side-leakage of lubricant cannot therefore be ignored. However, the general character of operation of a bearing is not greatly influenced by side flow and so may be most conveniently demonstrated by one-dimensional analysis, i.e. by the solution of equation 3.3.

On integrating equation 3.3 we obtain

$$\frac{dp}{dx} = 6\mu U\left(\frac{1}{h^2} - \frac{h_o}{h^3}\right) \tag{3.5}$$

in which $h = h_o$ where $dp/dx = 0$. The value of h_o and the position in the film where it occurs is as yet undetermined. On further integration

$$p - p_a = 6\mu U \int\left(\frac{1}{h^2} - \frac{h_o}{h^3}\right)dx$$

h_o and p_a are constants of integration and their values depend on the boundary conditions to be imposed and on the bearing geometry.

3.3.1 Wholly convergent film

If the film is wholly convergent, as is shown to an exaggerated scale in *Figure 3.3*, and the two ends of the film are exposed to the same ambient pressure, then the generated pressure distribution will build

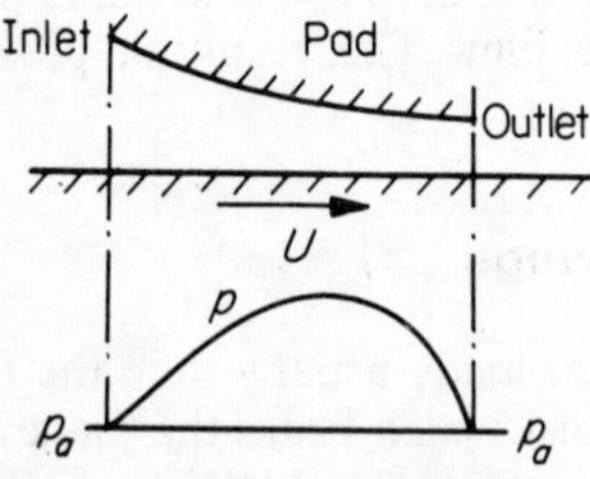

Figure 3.3. Wholly convergent film

up from the ambient value at the film inlet, rise to a maximum, and then fall back to the ambient value at the film outlet as shown. The generated pressure is everywhere positive within the length of the film.

3.3.2 Convergent-divergent film

Application of boundary conditions $p - p_a = 0$ at both ends of a convergent-divergent film produces a pressure distribution as shown in *Figure 3.4(a)*. Normally the ambient pressure p_a is small compared with the pressures generated within the film so that the predicted film pressure may become negative in the divergent region.

Whilst it is possible for liquids to sustain negative pressures, the presence of dissolved air in the liquid and the general condition of operation in bearings render negative pressures unlikely to occur. In fact there is ample evidence of cavitation occurring and of the pressure remaining at a near zero value in this region of the film.

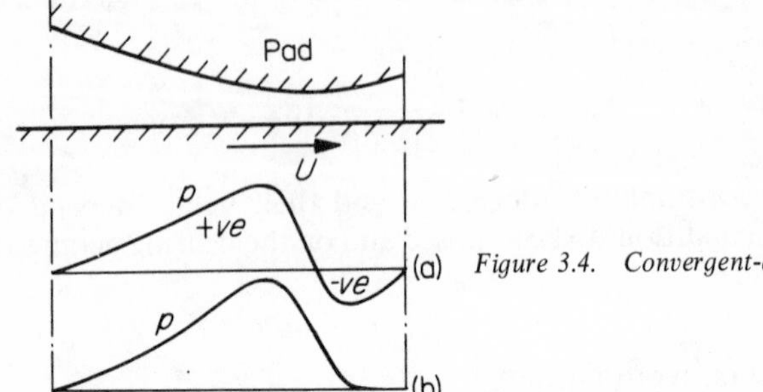

Figure 3.4. Convergent-divergent film

In a cavitated region equation 3.5 no longer applies and in consequence the use of a boundary condition at the outlet end of the film is invalid.

By considering continuity requirements it can be shown that the appropriate boundary condition is $p = 0$, $dp/dx = 0$, at some point within the divergent region of the film. The resulting pressure distribution is shown in *Figure 3.4(b)*.

3.4 Hydrodynamic thrust bearings

Bearings designed to carry shaft axial loads usually take the form of a ring of sector shaped pads against which bears the plane face of a collar (the runner) fixed to or integral with the shaft. With suitable provision of pad circumferential profile a lubricant film will be established between pad and runner when rotation occurs.

The precise form of the pad profile is not important provided the film is convergent i.e. the profile is such as to give decreasing film thickness in the direction of runner rotation. A particular case will be considered so as to indicate the procedure required in the analysis of thrust pad bearings.

3.4.1 Inclined pad—no side leakage

Consider a plane pad of length l inclined at a small angle α to a second plane surface which moves in its own plane with velocity U. It is convenient to use a coordinate system xy with its origin at the intersection of the two planes as shown in *Figure 3.5*. The film thickness at any point in the film is then simply $h = \alpha x$. Assuming that the film pressure varies only in the x direction (i.e. ignoring the effects of transverse or side-leakage flow) the appropriate starting point is equation 3.3

$$\frac{\mathrm{d}}{\mathrm{d}x}\left(h^3 \frac{\mathrm{d}p}{\mathrm{d}x}\right) = -6\mu U \frac{\mathrm{d}h}{\mathrm{d}x}$$

The right hand side of the equation is here negative because the runner velocity is in the negative direction as indicated in *Figure 3.5*.

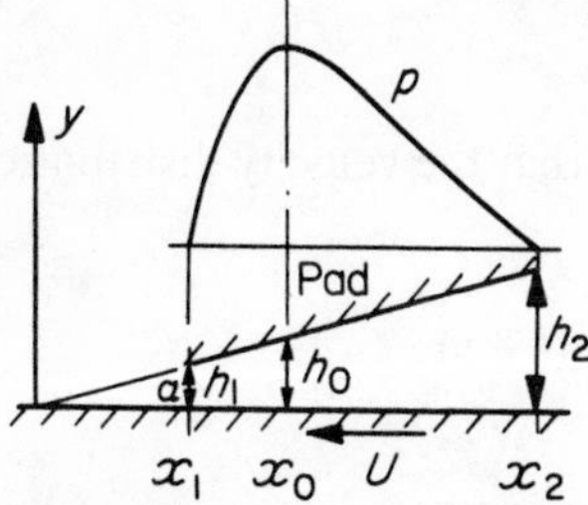

Figure 3.5. Notation for inclined pad

On integrating and substituting for h, and with $h_o = \alpha x_o$,

$$\frac{\mathrm{d}p}{\mathrm{d}x} = -\frac{6\mu U}{\alpha^2}\left(\frac{1}{x^2} - \frac{x_o}{x^3}\right)$$

and on further integration

$$p = \frac{6\mu U}{\alpha^2}\left(\frac{1}{x} - \frac{x_o}{2x^2} + C\right)$$

If the ambient pressure is taken as pressure datum, $p = 0$ at $x = x_1$ and x_2, i.e. at both ends of the pad then

$$x_o = \frac{2x_1x_2}{x_1+x_2}, \quad c = -\frac{1}{x_1+x_2}$$

$$\therefore p = \frac{6\mu U}{\alpha^2}\left\{\frac{(x_2-x)(x-x_1)}{x^2(x_1+x_2)}\right\}$$

The load supported per unit width of the pad is given by

$$P = \int_{x_1}^{x_2} p\,\mathrm{d}x = \frac{6\mu U}{\alpha^2}\left\{l_n \frac{x_2}{x_1} - \frac{2(x_2-x_1)}{(x_1+x_2)}\right\}$$

For a given pad inclination the position of the origin of x will depend upon the film thickness and, therefore, upon the load carried. It is therefore convenient to eliminate x_1 by introducing the film ratio k

where
$$k = \frac{h_2}{h_1} = \frac{x_2}{x_1}$$

whence
$$P = \frac{6\mu U}{\alpha^2}\left\{\ln k - \frac{2(k-1)}{(k+1)}\right\}$$

The viscous drag (friction) on the pad and runner surfaces are found by considering the shear stress in the film given by

$$\tau = \mu \frac{du}{dy}$$

In the present case $u = -U$ at $y = 0$ so that the velocity distribution within the film is

$$u = \frac{1}{2\mu}\frac{dp}{dx} y(y-h) - \frac{U}{h}(h-y)$$

and
$$\frac{dp}{dx} = -\frac{6\mu U}{\alpha^2}\left(\frac{1}{x^2} - \frac{x_o}{x^3}\right)$$

Hence
$$\tau = \frac{\mu U}{\alpha}\left\{\frac{1}{x} + \frac{3}{\alpha}\left(\frac{1}{x^2} - \frac{x_o}{x^3}\right)(\alpha x - 2y)\right\}$$

On the runner surface $y = 0$

$$\therefore \tau_o = \frac{\mu U}{\alpha}\left\{\frac{4}{x} - \frac{3x_o}{x^2}\right\}$$

On the pad surface $y = h$

$$\therefore \tau_h = \frac{\mu U}{\alpha}\left\{\frac{3x_o}{x^1} - \frac{2}{x}\right\}$$

Integration of the shear stress over the area of the pad gives the friction. Designating runner and pad friction per unit width of pad as F_o and F_h respectively

$$F_o = \int_{x_1}^{x_2} \tau_o \, dx = \frac{2\mu U}{\alpha}\left\{2\ln\frac{x_2}{x_1} - \frac{3(x_2 - x_1)}{(x_2 + x_1)}\right\}$$
$$= \frac{2\mu U}{\alpha}\left\{2\ln k - \frac{3(k-1)}{(k+1)}\right\}$$

and
$$F_h = \int_{x_1}^{x_2} \tau_h \, dx = \frac{2\mu U}{\alpha}\left\{\frac{3(k-1)}{(k+1)} - l_n k\right\}$$

It will be noticed that pad friction and runner friction are not equal. This is due to the fact that the film pressures produce a resultant force normal to the pad surface. Thus the total force on the pad in the direction of sliding is $F_h + \alpha P$ and this will be found to be equal to F_o, i.e.

$$F_o - F_h = \alpha P$$

A coefficient of friction f may be defined as the ratio of runner friction to load supported.

Thus $$f = \frac{F_o}{P} = \frac{\alpha}{3}\left\{\frac{2\ln k - [3(k-1)/(k+1)]}{\ln k - [2(k-1)/(k+1)]}\right\}$$

Since side-leakage has been ignored the lubricant flow rate is the same for all positions along the pad. At $x = x_o$ the pressure gradient is zero and the lubricant flow rate Q per unit width of pad is therefore simply

$$Q = \frac{Uh_o}{2}$$

but since $$h_o = \frac{2\alpha lk}{k^2-1}, \quad Q = U\alpha l\frac{k}{k^2-1}$$

The shearing of the lubricant gives rise to heat generation within the film. If it is assumed that no heat is transferred to either the pad or the runner the work done in shearing the film may be equated to the heat carried away in the lubricant in passing under the pad.

Thus $$F_o U = Q\rho c\Delta T$$

where c is the thermal capacity of the lubricant; ρ is the lubricant density; and ΔT is lubricant temperature rise.
Substituting for F_o and Q gives

$$\Delta T = \frac{\mu U}{J\rho c\alpha^2 l}\left\{\frac{4(k^2-1)\ln k - 6(k-1)^2}{k}\right\}$$

The above derived quantities may be written in terms of the minimum film thickness h_1 instead of pad inclination α by writing

$$\alpha = \frac{h_1}{l}(k-1)$$

Thus in non-dimensional form

$$\frac{Ph_1^2}{\mu U l^2} = 6\left\{\frac{\ln k}{(k-1)^2} - \frac{2}{k^2-1}\right\}$$

$$\frac{f \cdot l}{h_1} = \frac{k-1}{3}\left\{2 + \frac{k-1}{(k+1)\ln k - 2(k-1)}\right\}$$

$$\frac{\Delta T J \rho c h_1^2}{\mu U l} = \frac{2}{k}\left\{2\left(\frac{k+1}{k-1}\right)\ln k - 3\right\}$$

These quantities are plotted in *Figure 3.6* from which it will be seen that there is a maximum and a minimum respectively for load and coefficient of friction. This means that for given lubricant viscosity,

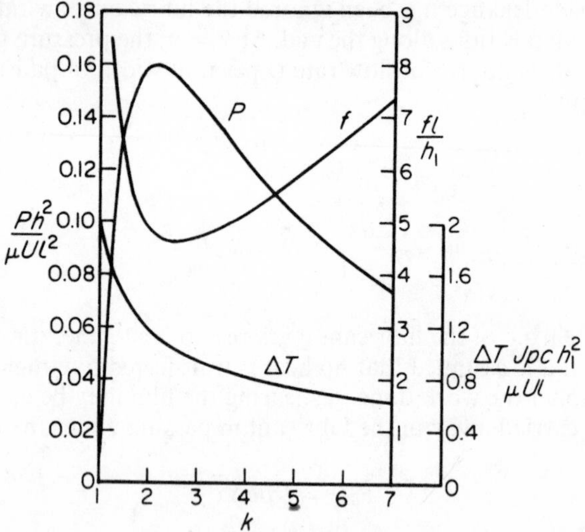

Figure 3.6. Performance characteristics for inclined pad—no side leakage

pad length and sliding velocity, a load will be supported with the largest film thickness if the pad inclination is such as to produce a film ratio k in the vicinity of 2. Approximately the same film ratio produces least coefficient of friction for a given operating film thickness. However, it will be evident from *Figure 3.6* that the value of film ratio is in no way critical since in the range 1.5 to 3.5 operation will be within about 20% of optimum conditions.

The minimum safe value of film thickness will depend on the surface finish of the pads and runner, the cleanliness of the lubricant, and the degree of perfection with which the bearing can be assembled.

Normally the operating film thickness in oil lubricated bearings is of the order of 0.02 mm. If a pad is to operate around optimum conditions it is necessary that the pad inclination should be such as to produce a variation of film thickness along the pad length also of the order of 0.02 mm. This necessarily involves a pad inclination which is numerically very small i.e. of the order of 10^{-3} for a pad 25 mm long and of smaller inclination for larger pads.

3.4.2 Tilting pads

If a pad is pivoted instead of being rigidly mounted the necessary inclination can be achieved automatically. It may be shown that the position of the centre of pressure on the pad face is given by

$$\frac{a}{l}=\frac{\frac{1}{2}(k^2-1)-(2k+1)\ln k+2(k-1)}{(k^2-1)\ln k-2(k-1)^2}$$

where a is the distance of the centre of pressure from the outlet end of the pad. If a pivot is provided at the back of the pad to coincide with the centre of pressure the value of k will be fixed for all operating conditions. For an optimum film ratio of about 2

$$\frac{a}{l}=0.42$$

For a centrally pivoted pad, i.e. $a/l=0.5$, simple theory predicts zero load carrying capacity. In fact centrally pivoted pads and, indeed, pads pivoted within the leading half of the pad are known to operate successfully but usually at a somewhat higher operating temperature. The ability of such pads to operate hydrodynamically can be attributed to two factors; the decrease of viscosity of the lubricant, as it progresses through the film, due to temperature rise thus causing a forward shift of the centre of pressure, and to thermal and bending distortion of the pad giving rise to a curved face to the pad.

Whilst tilting pads obviate the need for close tolerance machining of the pads to produce a profile and, incidentally, permit operation in either direction of rotation, a difficulty arises in the provision of pivots to present the pad faces uniformly to the face of the runner. Any error in pad positioning is reflected in a variation in film thickness from one pad to the next and, due to the extreme sensitivity of load capacity to film thickness, a high order of accuracy of pad mounting is necessary in order to ensure uniform distribution of load among the pads. If, in a bearing comprising a number of pads designed to operate with a film thickness of 0.02 mm, one pad is low by 0.004 mm compared with the remaining pads, this one pad will carry only about 65% of its share of the load. Such maldistribution of load in a bearing is inefficient and can lead to excessive friction and to bearing failure in extreme cases. A bearing ring with integral machined pads is also subject to unequal loading of pads if the pads are of different height but for small bearings it is not difficult to ensure equality of height to a high order of accuracy.

3.4.3 Pads of finite width

Practical bearing pads are subject to the ambient pressure at the sides of the pad as well as at the leading and trailing edges. This results in an approximately parabolic pressure distribution across

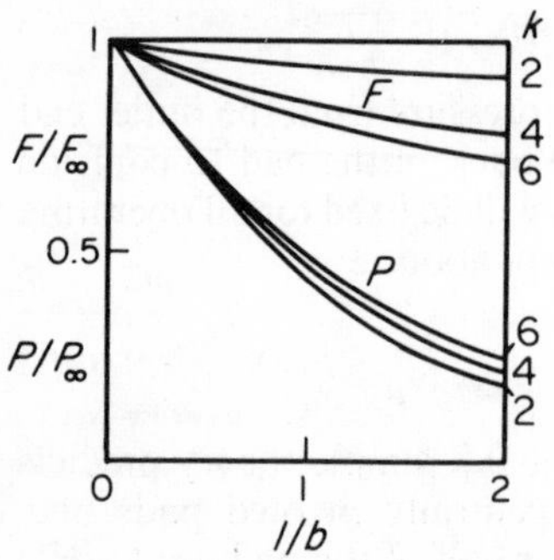

Figure 3.7. Leakage factors for inclined pads. P and F, actual load capacity and friction P_∞ and F_∞, load capacity and friction calculated on basis of no-side leakage

the width of the pad and an accompanying side flow of lubricant from both side edges of the pad. The principal effect of this side-leakage on the performance of the pad is that the load capacity predicted by the no-side-leakage theory is appreciably reduced. In *Figure 3.7* the load capacity per unit width is compared with the no-side-leakage case. The reduction in load capacity is principally a function of the length to width ratio of the pad. It is not much affected by the film ratio but the reduction decreases for large values of film ratio. It is evident that for a square pad, i.e. $l/b = 1$, the load capacity is rather less than half that predicted on the basis of no-side leakage. The dependence of total bearing load capacity upon both length and width of pads results in square pads approximating to optimum pad proportions.

Pad friction is not greatly affected by the side leakage since the velocity gradient within the film is only partly dependent upon the pressure gradient in the film. For a square pad the friction is 94%

of the value predicted on the basis of no-side-leakage theory for a film ratio of 2.

In a practical bearing the prediction of operating temperature is complicated not only by the question of heat transfer from lubricant to pads but also by the side flow of lubricant leaving the pad at temperatures varying from that at pad inlet to that at the trailing edge.

For pivoted pads the optimum position of the pivot is very little affected by side-leakage and the position indicated by the simple theory is suitable for pads of all practical proportions.

3.5 Hydrodynamic journal bearings

As a class, journal bearings differ significantly from thrust bearings and therefore merit separate consideration. In the first place, the function of the journal bearing is to provide radial, and therefore two-dimensional location for a journal as distinct from the one-dimensional axial location provided by a thrust bearing. Secondly, for a simple cylindrical journal bearing the necessary film profile arises naturally as the journal moves off-centre with respect to the bearing and therefore does not require special provision of pad profile as is the case in a thrust bearing.

It is necessary to differentiate between 'clearance' and 'fitted' bearings. A bearing subtending an arc greater than 180° must necessarily be a clearance bearing, i.e. journal diameter must be less than bearing diameter, whereas a partial bearing subtending less than 180° could have curvature equal to that of the journal. A 360° bearing is clearly capable of carrying a journal load in any direction normal to the journal axis. A partial bearing, on the other hand, is limited to loads which fall within the arc subtended by the bearing.

3.5.1 360° journal bearing—no side leakage

Consider a journal of radius R within a bearing of radius $R+c$ as shown in *Figure 3.8*. The radial clearance c is very small compared with R, c/R being usually of the order of 10^{-3}. Generally the journal centre J will be displaced from the bearing centre B in operation by an eccentricity e. If the line through J and B defines the origin of an angular coordinate θ, then the film thickness h between journal and bearing will be given by

$$h = c + e\cos\theta,$$

since c/R is small, or

$$h = c(1+\varepsilon\cos\theta)$$

where eccentricity ratio $\varepsilon = e/c$.

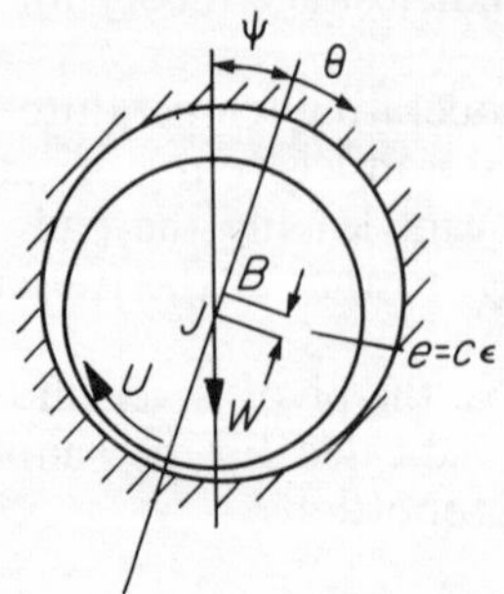

Figure 3.8. Notation for 360° journal bearing

Substituting for $x = R\theta$ and for h in equation 3.5

$$\frac{\mathrm{d}p}{\mathrm{d}\theta} = \frac{6\mu UR}{c^2}\left\{\frac{1}{(1+\varepsilon\cos\theta)^2} - \frac{h_o}{c}\,\frac{1}{(1+\varepsilon\cos\theta)^3}\right\}$$

To integrate further let

$$1+\varepsilon\cos\theta = \frac{1-\varepsilon^2}{1-\varepsilon\cos\phi}$$

from which it may be deduced that

$$\cos\theta = \frac{\cos\phi-\varepsilon}{1-\varepsilon\cos\phi}$$

$$\sin\theta = \frac{(1-\varepsilon^2)^{1/2}\sin\phi}{1-\varepsilon\cos\phi}$$

and

$$\frac{\mathrm{d}\theta}{\mathrm{d}\phi} = \frac{(1-\varepsilon^2)^{1/2}}{1-\varepsilon\cos\phi}$$

On substitution

$$\frac{\mathrm{d}p}{\mathrm{d}\phi} = \frac{6\mu UR}{c^2}\left\{\frac{1-\varepsilon\cos\phi}{(1-\varepsilon^2)^{3/2}} - \frac{h_o}{c}\,\frac{(1-\varepsilon\cos\phi)^2}{(1-\varepsilon^2)^{5/2}}\right\}$$

If the lubricant is fed to the bearing at $\theta = 0$ and at pressure negligible compared with mean bearing pressure, we may adopt as boundary conditions for the oil film

$$p = 0 \text{ at } \theta = 0, \text{ i.e. at } \phi = 0$$

and

$$p = \frac{\mathrm{d}p}{\mathrm{d}\phi} = 0 \text{ at } \theta = \theta_2, \text{ i.e. } \phi = \phi_2$$

whence
$$\frac{h_o}{c} = \frac{1-\varepsilon^2}{1-\varepsilon\cos\phi_2}$$

where ϕ_2 is given by

$$\varepsilon = -2\left(\frac{\phi_2\cos\phi_2-\sin\phi_2}{\phi_2-\sin\phi_2\cos\phi_2}\right)$$

and

$$p = \frac{6\mu UR}{c^2(1-\varepsilon^2)^{3/2}}\left\{\phi-\varepsilon\sin\phi-\frac{(2+\varepsilon^2)\phi-4\varepsilon\sin\phi+\varepsilon^2\sin\phi\cos\phi}{2(1-\varepsilon\cos\phi_2)}\right\}$$

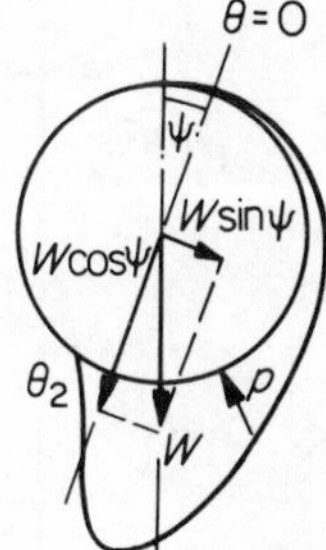

Figure 3.9. Forces acting upon journal

The form of this pressure distribution around the journal is illustrated in *Figure 3.9*.

Consider now the equilibrium of the journal under the action of an external force W and the pressure p acting normal to the journal surface over the width b of the bearing. The effect of lubricant shear stress acting on the journal surface will be small and is therefore neglected.

Resolving along the line of centres

$$W\cos\psi = -b\int_0^{\theta_2} p\cos\theta\cdot R\,d\theta$$

and perpendicular to the line of centres

$$W\sin\psi = b\int_0^{\theta_2} p\sin\theta\cdot R\,d\theta$$

whence,

$$W = \frac{6\mu U R^2 b}{c^2} \frac{1}{(1-\varepsilon^2)^{1/2}(1-\varepsilon\cos\phi_2)} \left\{(\phi_2\cos\phi_2 - \sin\phi_2)^2 + \frac{\varepsilon^2(1-\cos\phi_2)^4}{4(1-\varepsilon^2)}\right\}^{1/2}$$

and,

$$\tan\psi = -\frac{2(1-\varepsilon^2)^{1/2}(\phi_2\cos\phi_2 - \sin\phi_2)}{\varepsilon(1-\cos\phi_2)^2}$$

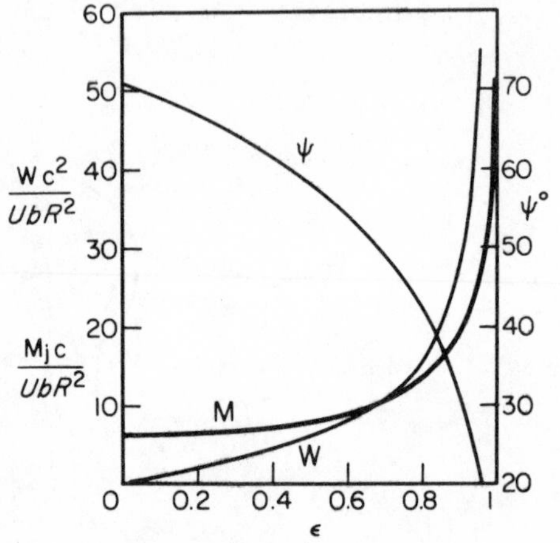

Figure 3.10. Performance characteristics for 360° journal bearing—no side leakage

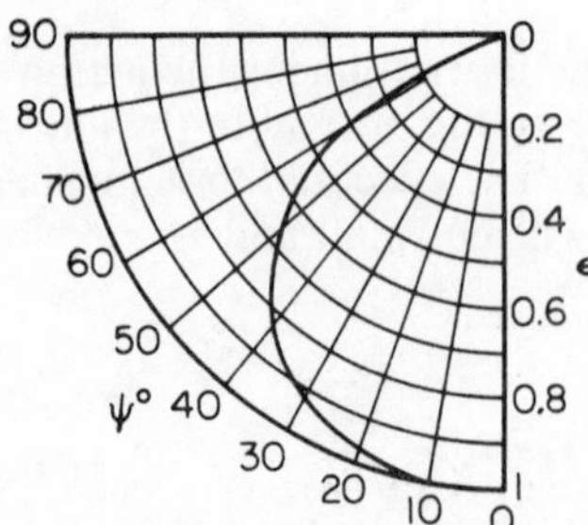

Figure 3.11. Journal centre locus for 360° journal bearing—no side leakage

The load, non-dimensionalised in the form $(Wc^2)/(\mu U R^2 b)$ and the attitude angle ψ are given in terms of the eccentricity ratio in *Figure 3.10*. The operating position of the journal within the bearing can best be seen from a polar plot of eccentricity ratio against attitude angle as shown in *Figure 3.11*. For otherwise constant operating conditions, increase of load causes the journal centre to follow a roughly semi-circular path from a concentric position at no load towards a condition of contact between journal surface and bearing at a point in the direction of the applied load at heavy loads.

3.5.2 Journal friction

The shear stress acting upon the journal surface is given by

$$\tau_j = \mu\left(\frac{\mathrm{d}u}{\mathrm{d}y}\right)_j$$

The velocity distribution within the film is given by:

$$u = \frac{U}{h}(h-y) + \frac{1}{2\mu R}\frac{\mathrm{d}p}{\mathrm{d}\theta}y(y-h)$$

where $y = 0$ corresponds to the journal surface.

Whence, $$\tau_j = -\frac{\mu U}{h} - \frac{h}{2R}\frac{\mathrm{d}p}{\mathrm{d}\theta}$$

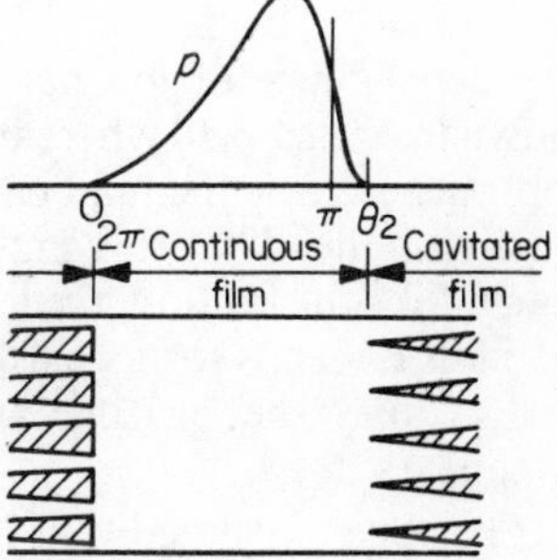

Figure 3.12. Continuous and cavitated film regions in journal bearing

From $\theta = 0$ to θ_2 the film is continuous, but from θ_2 to 2π the film is cavitated and the pressure gradient is zero. By virtue of the assumption that there is no side leakage flow of lubricant from the film, the quantity of lubricant flowing per unit width of the bearing is constant at $(Uh_o)/2$. In consequence in the region θ_2 to 2π in which the film thickness is greater than h_o the film is striated, the flow breaking into a number of narrow streams in the vicinity of θ_2 and reforming again at 2π as illustrated in *Figure 3.12*. The effective width b' of the cavitated region is given by

$$b'h = bh_o$$

i.e.
$$b' = b\frac{1+\varepsilon\cos\theta_2}{1+\varepsilon\cos\theta}$$

Hence, frictional moment on the journal

$$M_j = R^2\left\{b\int_0^{\theta_2}\left(\frac{\mu U}{h}+\frac{h}{2R}\frac{dp}{d\theta}\right)d\theta+\int_{\theta_2}^{2\pi} b'\frac{\mu U}{h}d\theta\right\}$$

$$= R^2 b\left[\frac{\mu U}{c}\int_0^{\theta_2}\frac{d\theta}{1+\varepsilon\cos\theta}+\frac{c}{2R}\int_0^{\theta_2}\left\{(1+\varepsilon\cos\theta)\frac{dp}{d\theta}\right\}d\theta\right.$$
$$\left.+\frac{\mu U}{c}\int_{\theta_2}^{2\pi}\frac{(1+\varepsilon\cos\theta_2)}{(1+\varepsilon\cos\theta)^2}d\theta\right]$$

On substituting for $dp/d\theta$ and for θ in terms of ϕ the integration is straight forward and gives

$$M_j = \frac{R^2 b\mu U}{c}\left\{\frac{2\pi+4\varepsilon\sin\phi_2-4\phi_2\varepsilon\cos\phi_2}{(1-\varepsilon^2)^{1/2}(1-\varepsilon\cos\phi_2)}\right\}$$

The dependence of journal friction upon the eccentricity ratio is shown in *Figure 3.10* where it will be noticed that as the eccentricity approaches zero, the load capacity tends to zero whilst the friction remains finite. That this must be so will be obvious from the fact that, unlike the friction, the load carrying pressure generation within the films depends for its existence upon the variation of film thickness around the bearing and at zero eccentricity the variation disappears.

It has been assumed that the film pressure is at the datum value $p = 0$, at $\theta = 0$, i.e. on the line of centres the position of which is given by the attitude angle ψ. In a real bearing the datum pressure is defined in value and location by an oil feed hole or groove at a fixed position. *Figures 3.10* and *3.11* show how the attitude angle varies with the eccentricity ratio, and an oil feed hole correctly placed for a particular operating condition will be incorrectly placed for another operating condition. In fact the predicted load capacity of the bearing is not greatly affected by variation of the oil feed position provided that it is placed in the unloaded half of the bearing. If the feed hole were placed in the high pressure region of the bearing, pressure generation within the film would be grossly interfered with unless a correspondingly high feed pressure were used.

3.5.3 Effects of side leakage

In practice, the majority of journal bearings have width to diameter ratio in the range $\frac{1}{4}$ to $1\frac{1}{2}$ and the side-leakage flow of lubricant results in an appreciable reduction of load capacity and a small reduction in friction for a given eccentricity ratio. The load capacity per unit width and friction compared with the no-side-leakage case is shown for a number of values of eccentricity ratio in *Figure 3.13*.

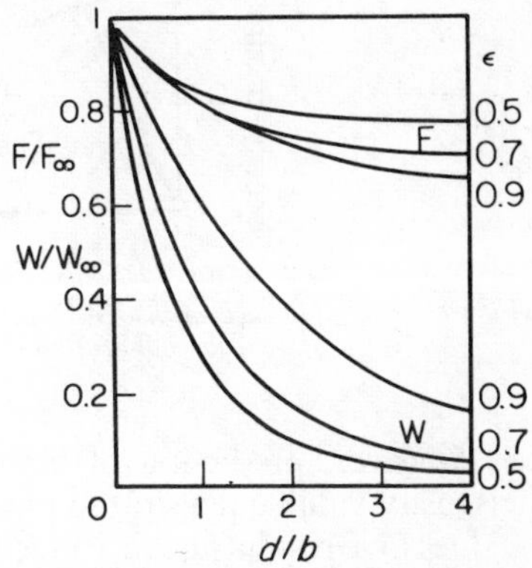

Figure 3.13. Leakage factors for 360° journal bearing. W and F actual load capacity and friction W_∞ and F_∞, load capacity and friction calculated on basis of no side leakage

In addition to the effects mentioned above, side-leakage plays a fundamentally important role in the 360° journal bearing as this is the only way in which the lubricant can leave the bearing. In the absence of side-leakage flow, the heat generated within the film would necessarily have to be conducted away through the bearing wall and other machine members. Such a mechanism would be satisfactory at low speed but would otherwise involve unacceptably high operating temperatures. The proportion of the total energy dissipated that is carried away by the through flow of lubricant varies considerably from case to case but for medium and high speed bearings is typically between 0.7 and 1.0.

3.5.4 Effective film viscosity

Due to the inevitable temperature rise of the lubricant as it passes along the film in a bearing there is a consequent variation in the lubricant viscosity. For normal oil lubricants the viscosity can fall by 50% for a temperature rise of between 10 °C and 30 °C so that a prediction of bearing performance based on lubricant viscosity at oil entry temperature can be greatly in error.

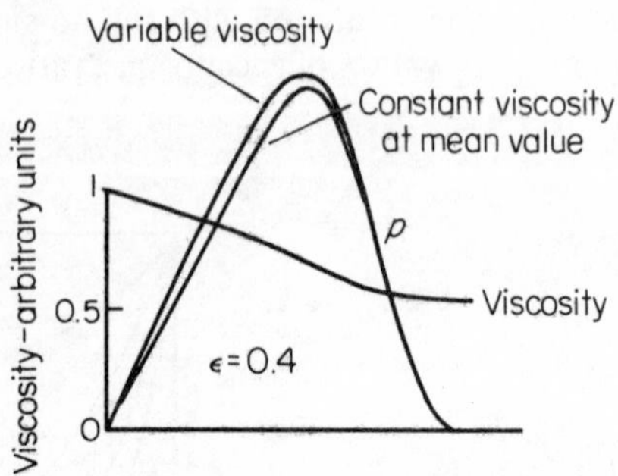

Figure 3.14. Influence of viscosity variation on calculated pressure distribution in journal bearing—no side leakage theory

However, prediction of film pressure distribution based on a viscosity value equal to the mean of film maximum and minimum values give results not very much different from those obtained with due allowance for point to point variation in viscosity. This is illustrated for a particular case in *Figure 3.14*. It may seem surprising that the variation in viscosity does not result in a more pronounced change in the pressure distribution. The reason lies in the fact that the pressure at any particular point is determined by the viscosity at every point along the film and not especially by the viscosity at the particular point.

3.6 Hydrostatic lubrication

Consider a pair of surfaces, one finite in extent, separated by a lubricant film of uniform thickness h. The pressure distribution over the finite surface is given by equation 3.4, i.e.

$$\frac{\partial^2 p}{\partial x^2} + \frac{\partial^2 p}{\partial z^2} = 0 \tag{3.6}$$

At the perimeter of the finite surface the pressure will be that of the surroundings which for convenience may be taken as the pressure datum. If a lubricant source at pressure p_1 exists somewhere within the film the pressure distribution may be found in terms of p_1.

For simplicity consider the case of a circular pad of radius R_2 having a concentric recess of radius R_1 as shown in *Figure 3.15*.

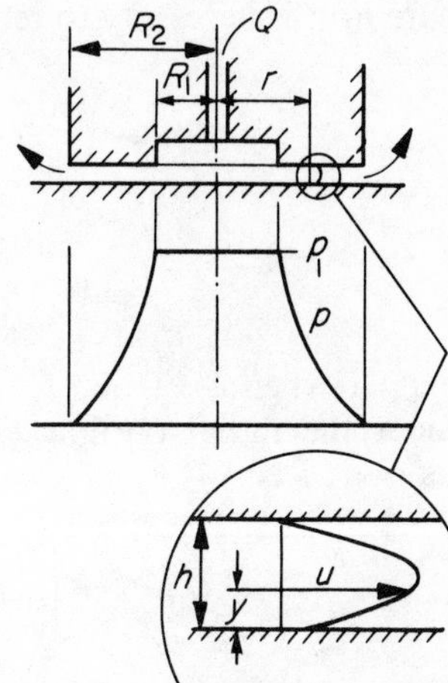

Figure 3.15. *Notation for circular hydrostatic pad*

For this axisymmetric case equation 3.6 is better written in polar form

$$\frac{1}{r}\frac{\mathrm{d}}{\mathrm{d}r}\left(r\frac{\mathrm{d}p}{\mathrm{d}r}\right) = 0$$

On integration,

$$p = c_1 \ln r + c_2$$

Provided that the recess is deep compared with the thickness h of the film we may assume that the pressure is constant at p_1 over the entire recess area.

With boundary conditions

$$p = 0,\ r = R_2$$

and
$$p = p_1,\ r = R_1$$

$$c_1 = \frac{p_1}{\ln(R_1/R_2)}$$

and
$$c_2 = \frac{-p_1 \ln R_2}{\ln(R_1/R_2)}$$

whence
$$p = p_1 \frac{\ln(r/R_2)}{\ln(R_1/R_2)}$$

and the pressure distribution is as indicated in *Figure 3.15*.

Further information than this is not available from equation 3.6. To determine the lubricant flow rate necessary to maintain the recess pressure p_1 it is necessary to return to first principles,

$$\frac{\mathrm{d}p}{\mathrm{d}r} = \mu \frac{\partial^2 u}{\partial y^2}$$

$$\therefore u = \frac{1}{2\mu} \frac{\mathrm{d}p}{\mathrm{d}r} y(y-h)$$

since $u = 0$ when $y = 0$ and h.

Consider the radial outflow Q across any cylindrical surface of radius r

$$Q = 2\pi r \int_0^h u \,\mathrm{d}y = -\frac{\pi r h^3}{6\mu} \frac{\mathrm{d}p}{\mathrm{d}r}$$

$$\therefore \frac{\mathrm{d}p}{\mathrm{d}r} = -\frac{6\mu Q}{\pi h^3 r}$$

Integrating with respect to r and noting that $p = 0$ at $r = R_2$

$$p = -\frac{6\mu Q}{\pi h^3} \ln \frac{r}{R_2}$$

At $r = R_1$,
$$p = p_1 = \frac{6\mu Q}{\pi h^3} \ln \frac{R_2}{R_1}$$

The load supported by the pad is obtained by integrating the pressure over the pad area, giving

$$w = \frac{\pi}{2} \frac{R_2^2 - R_1^2}{\ln(R_2/R_1)} p_1$$

More generally,
$$p_r = \Phi \mu Q / h^3$$

and
$$w = \Psi A p_r$$

where p_r is recess pressure;
A is total pad area;
Φ and Ψ are functions only of pad geometry.

If a bearing comprises a number of pads connected to a common lubricant supply it is usual to introduce flow resistance between the supply and the individual pads to prevent premature shut down of any one pad. Such an arrangement is shown diagrammatically in *Figure 3.16*. If one pad lifts more than the others the flow to that pad increases. The increased flow causes increased pressure drop in the flow restrictor associated with that pad and the recess pressure

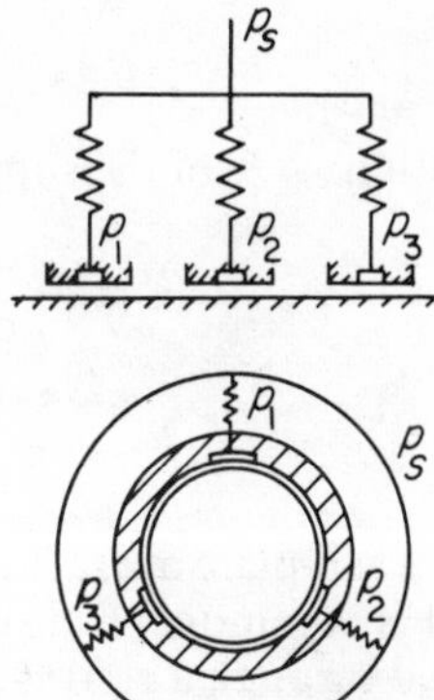

Figure 3.16. Multi-pad hydrostatic bearings

of that pad decreases. The external load on the pad causes the film thickness to decrease and the situation is stabilised. A pad shuts down when the flow falls to zero. There is then no pressure drop in the restrictor and the recess pressure equals the supply pressure. Since the supply pressure is common to all pads of a bearing the shut down load for each pad is simply given by

$$W_{\text{Shut down}} = \Psi A p_s \quad \text{where } p_s \text{ is supply pressure.}$$

A further advantage of the use of flow restrictors arises from the control afforded over the effective stiffness of the oil film. For the pad we may define a flow impedance z_1 in terms of the pressure drop p_1 through the pad and the flow Q.

$$z_1 = \frac{p_r}{Q} = \Phi\mu/h^3$$

For a capillary restrictor we may similarly define a flow impedance

$$z_0 = \frac{p_s - p_r}{Q}$$

(For an orifice restrictor the pressure drop is not linearly related to the flow rate.)

In terms of the two impedances we obtain

$$p_r = \frac{z_1}{z_1+z_0} p_s$$

whence, pad load

$$w = \Psi A \frac{z_1}{z_1+z_0} p_s$$

$$= \frac{\Psi A p_s}{1+(z_0 h^3/\mu\Phi)}$$

The stiffness K of the supporting film is given by

$$K = -\frac{\mathrm{d}w}{\mathrm{d}h} = \frac{\Psi A p_s(3h^2 z_0/\mu\Phi)}{\{1+(z_0 h^3/\mu\Phi)\}^2}$$

$$\therefore K = \frac{3\Psi A p_s(z_0/z_1)}{h\{1+(z_0/z_1)\}^2} = \frac{3w}{h\{1+(z_1/z_0)\}}$$

In swash-plate operated axial piston pumps and motors the piston thrust is commonly transmitted by a circular pad directly supplied by fluid through a restrictor in the piston. By suitable choice of pad dimensions in relation to the piston area, direct hydrostatic lubrication is achieved. The volumetric efficiency of the pump or motor is of course reduced since some of the fluid handled is diverted to the pads and an analysis of this condition is presented in Chapter 5.

Radial support of a journal may be achieved hydrostatically by providing three or more recesses distributed around a journal bearing. In such an arrangement the individually restricted recesses act in opposition and, since the forces exerted by the separate films are not limited to the external journal load, a high stiffness of journal support can be achieved.

The principal advantage of hydrostatic lubrication is that it is not dependent upon relative motion of the opposing surfaces for its action. An effective separating film exists even under starting and stopping conditions which under normal hydrodynamic lubrication would involve contact between opposing surfaces.

BIBLIOGRAPHY

Shaw and Macks, *Analysis and Lubrication of Bearings*, McGraw-Hill (1949).
Barwell, *Lubrication of Bearings*, Butterworths, (1956).
Wilcock, D. F. and Booser, E. R., *Bearing Design and Application*, McGraw-Hill, (1957).
Hersey, *Theory and Research in Lubrication*, Wiley, (1966).
Pinkus, O. and Sternlicht, B., *Theory of Hydrodynamic Lubrication*, McGraw-Hill, (1961).

Chapter 4

Fluid Seals

4.1 Introduction

The function of a fluid seal is to limit leakage between various parts of a machine, usually parts in relative motion. In practice it is rarely possible to achieve a leaktight seal between moving parts and the seal designer has to achieve a compromise between the incompatible demands of low leakage on the one hand and low rates of wear, friction and power loss on the other. There are frequently special difficulties where the seal environment is severe, due perhaps to extremes of temperature or pressure, to high sliding speeds or to fluids which are incompatible with the materials of the seal.

For the majority of dynamic seals a working life of several thousand hours is required, such seals, whether reciprocating or rotary, are usually designed to function so that the sliding faces are lubricated by the medium being sealed. Lubrication reduces wear and friction and very low leakage levels can still be achieved. Measurements show that the lubricating film may often be no more than a micron or two in thickness and all the evidence points to the existence of a hydrodynamic effect which produces and maintains this vital film of lubricant. The theoretical basis of seal lubrication is similar to that of bearings, which often have similar configurations, although higher rates of flow are usual in bearing films. The precise details of the mechanism which allows the hydrodynamic effect to be produced in seals are not well understood but it is generally agreed that the profiles of the sliding faces of the seal are all-important. In some seals the elastic and other properties of the contact face materials have also to be considered.

A resumé of the principal types of dynamic seals is given in *Table 4.1a* and *Table 4.1b* summarises some of the important variables.

Table 4.1a

Rotary seals		*Reciprocating seals*	
Rigid	*Flexible*	*Rigid*	*Flexible*
Face seal* Bush seal Labyrinth Screw-seal Centrifugal seal Hydrostatic seal	Stuffing box Lip-seal O-ring Felt ring	Bush seal Piston-ring	Stuffing box Square-back U-ring V-ring O-ring X-ring Bellows Diaphragm

* Also called by other names such as: mechanical seal, radial face seal, carbon face seal, etc.

Table 4.1b

Independent variables		*Dependent variables*	
Seal properties	*Operating conditions*	*Primary*	*Secondary*
Geometry Face profile Materials	Sealed pressure Sliding speed Closing force Fluid: Viscosity Thermal conductivity Density etc.	Leakage Friction Wear-rate	Face separation Hydrodynamic pressure Viscosity Film temperature Seal temperature Surface tension

4.2 Seal lubrication theory

Since the lubricant films in seals are usually very thin, the Reynolds numbers involved are usually very low. Therefore viscous shear and fluid pressure are the only significant forces to be considered. Flow across the thickness of the film is negligible and the pressure may therefore be taken to be constant in this direction. In comparison with $\partial u/\partial y$ and $\partial w/\partial y$, all other velocity gradients are negligible (*Figure 4.1* indicates the notation). With these assumptions the equations of motion of the fluid in the film reduce from the full Navier–Stokes equations discussed in Chapter 1 and may be written as

$$\frac{\partial^2 u}{\partial y^2} = \frac{1}{\mu}\frac{\partial p}{\partial x} \quad \text{and} \quad \frac{\partial^2 w}{\partial y^2} = \frac{1}{\mu}\frac{\partial p}{\partial z} \qquad (4.1) \text{ and } (4.2)$$

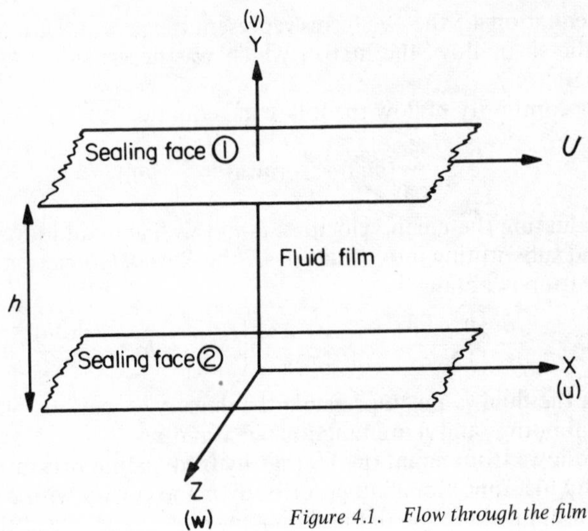

Figure 4.1. Flow through the film

When the relative motion of the bounding surfaces is parallel to x, equations 4.1 and 4.2 may be integrated across the film using the boundary conditions:

$$w = 0,\ u = 0;\ y = 0$$

and

$$w = 0,\ u = U;\ y = h$$

hence the velocity components of the fluid are:

$$u = \frac{1}{2\mu} \cdot \frac{\partial p}{\partial x} y(y-h) + \frac{Uy}{h} \tag{4.3}$$

$$w = \frac{1}{2\mu} \frac{\partial p}{\partial z} y(y-h) \tag{4.4}$$

The slight difference between the equation for u and the one in Chapter 3 (page 47) is due to the interchange of the values of u at $y = 0$ and $y = h$.

4.2.1 Flow rates and Reynolds equation

By integrating equations 4.3 and 4.4 with respect to y, the flow rates in each direction are obtained:

$$q_x = -\frac{h^3}{12\mu} \frac{\partial p}{\partial x} + \frac{Uh}{2} \quad \text{per unit length} \tag{4.5}$$

$$q_z = -\frac{h^3}{12\mu} \frac{\partial p}{\partial z} \quad \text{per unit length} \tag{4.6}$$

In equation 4.5 the two terms represent the pressure gradient flow and the shear flow, the first of which was described in Chapter 1 (page 4).

For continuity of flow the following equation must be fulfilled

$$\frac{\partial}{\partial x}(\rho\bar{u}h)+\frac{\partial}{\partial z}(\rho\bar{w}h) = \frac{\partial}{\partial t}(\rho h) \tag{4.7}$$

Evaluating the mean velocities, $\bar{u}$ and $\bar{w}$, from equations 4.3 and 4.4 and substituting into equation 4.7 the *Reynolds equation* for fluid lubrication is obtained

$$\frac{\partial}{\partial x}\left\{\frac{h^3\rho}{12\mu}\frac{\partial p}{\partial x}\right\}+\frac{\partial}{\partial z}\left\{\frac{h^3\rho}{12\mu}\frac{\partial p}{\partial z}\right\} = \frac{\partial}{\partial t}(\rho h)+\frac{u}{2}\frac{\partial(\rho h)}{\partial x} \tag{4.8}$$

When the fluid is incompressible the density, ρ, cancels out but in general both ρ and μ are functions of x and z.

It follows from equation 4.8 that hydrodynamic pressure can be built up in a thin film if either, or both, h and ρ vary with respect to time or in the direction of the sliding motion, i.e. provided the terms on the right hand side of the equation do not both vanish. In terms of a 'rigid' seal, this could come about through the sliding surfaces not being completely flat and parallel, as is the case when there is residual waviness or misalignment of the order of the film thickness.

Where the surfaces are deformable, in a rubber seal for instance, then the Reynolds equation has to be solved simultaneously with the equations of elasticity. In practice a lack of knowledge about the form of $h(x)$ prevents the Reynolds equation being solved even for rigid surfaces.

4.2.2 Friction

The friction due to viscous drag is obtained by differentiating equations 4.3 and 4.4 with respect to y and substituting for the velocity derivatives in the equation which defines Newtonian viscosity:

$$\tau = \mu\frac{\partial u}{\partial y} \tag{4.9}$$

Since the pressure gradient flows are usually small compared with the flow due to the drag of the moving surface only the second term on the right hand side of equation 4.3 is important in this connection and the friction force is therefore:

$$\tau = \frac{\mu U}{h} \quad \text{per unit area} \tag{4.10}$$

4.2.3 Non-dimensional parameters

When equation 4.8 is written in non-dimensional form for an incompressible fluid it becomes:

$$\frac{\partial}{\partial X}\left(\frac{H^3}{M}\frac{\partial P}{\partial X}\right)+\frac{\partial}{\partial Z}\left(\frac{H^3}{M}\frac{\partial P}{\partial Z}\right)=A\frac{\partial H}{\partial T}+B\frac{\partial H}{\partial X} \tag{4.11}$$

where $x = x_0X$; $z = x_0Z$; $t = t_0T$; $h = h_0H$; $\mu = \mu_0M$ and $p = p_0P$

and $$A=\frac{12\mu_0 x_0^2}{h_0^2 p_0 t_0}, \quad B=\frac{6\mu_0 U x_0}{h_0^2 P_0}$$

The non-dimensional constants are the 'Squeeze Number' (A) and the 'Bearing Number' (B). Appropriate quantities (suffix zero) are selected for converting the variables to non-dimensional form.

The bearing number is of particular importance in determining the average hydrodynamic pressure in a thin film, it is also a measure of the relative importance of the shear flow and pressure gradient flow in the film, this may be seen by considering the ratio of the terms in equation 4.5. In general, the larger the bearing number the larger is the hydrodynamic pressure generated in the film.

The friction coefficient for a seal film may be derived from equation 4.10 by dividing by the load per unit area (W):

$$\text{friction coefficient, } f=\frac{\mu U}{hW} \tag{4.12}$$

The group $\mu U/W$ has the dimensions of $(\text{length})^{-1}$ and is useful as an indication of the severity of the operating conditions of a seal. When this 'Duty Parameter' (G) is large the seal is expected to have a thick film and be working in the hydrodynamic lubrication regime and when it is small, in the region of boundary lubrication. It is important that the value of the viscosity used should be that pertaining to the film, rather than the bulk value for the sealed fluid, which may be quite different. It is sometimes the practice to convert the duty parameter into non-dimensional form (G') by multiplying by the width of the sealing face, then if D is the mean diameter of the face, the ratio $(\pi DG')/f$ is equal to the mean film thickness as indicated in *Figure 4.4*.

4.2.4 Boundary lubrication

When operating under arduous conditions, at low values of the duty parameter, the sliding faces of the seal begin to make solid contact at asperities which penetrate the hydrodynamic film, this marks the transition to boundary lubrication. In this regime the wear and friction are very dependent upon the physical chemistry of the sliding surfaces. Good boundary lubricants have a facility for shearing easily and preventing the formation of strong bonds or welds between the two surfaces. Phosphates, sulphides, chlorides and oxides of metals often behave in this way. Structural materials which are good boundary lubricants include graphite carbon (in the presence of oxygen or water vapour), tungsten carbide, alumina-based ceramics and the plastic material polytetrafluoroethylene (ptfe).

4.3 Materials for seals

It is important that the structural materials of a seal should be compatible with the fluid to be sealed and should be able to withstand the temperature to which they will be subjected. These requirements are of particular importance in the case of rubber seals. For mechanical seals an additional requirement is that the contact face materials should be good bearing materials, very often too they must operate under adverse conditions where solids in suspension create a serious wear problem.

4.3.1 Polymers for seals

Rubbers (elastomers) and plastics are based on long chain polymers but in rubber the individual molecules are folded and cross-linked so that the material shows elastic properties which are lacking in plastics. When a rubber is subjected to excessive stress, or temperature, or is exposed to chemically active materials, the cross links may break and reform so that the material is permanently deformed (permanent set). In a practical seal design this possibility must be minimised since the accompanying stress relaxation may reduce the seal contact pressure to a level where it no longer seals effectively. Atmospheric oxygen, ozone and water have been shown to affect certain rubbers in this way.

Absorption of fluids by some rubbers is also a problem and care must be taken in selecting a polymer base for a rubber to ensure fluid compatibility. Absorption is usually accompanied by swelling and general weakening of the rubber, leading to early seal failure.

Within a broad class of rubbers based on a particular polymer, significant variations in the detailed properties are found, depending

Table 4.2 PROPERTIES OF SEAL ELASTOMERS

	Natural rubber	*Butyl*	*SBR*	*Neoprene*	*Nitrile*	*Poly-acrylate*	*Poly-sulphide*	*Poly-urethane*	*Silicone*	*Fluoro-silicone*	*Fluoro-carbon*	*Ethylene-propylene*	*PTFE (plastic)*	*Polyimide (plastic)*
Approximate, max.	80	110	.	130	145	160	105	135	250	175	225	150	280	300
Temp. range, min.	−55	−60	−55	−35	−45	−30	−55	−55	−95	−75	−40	−55	−100	−250
Durability	**	=	*	**	*	=	−	**	−	−	*	.	−	**
Resilience	**	−	=	*	=	=	−	*	*	.	=	.	−	−
Compression set	**	−	.	*	**	−		*	*	.	*	=	−	**
Oil resistance	−	−	−	=	**		**	*	=	*	**	.	**	**
Water resistance	**	*	**	=	=	=		$(=)^1$	*	*	$(=)^2$	.	**	$(=)^3$
Phosphate ester resistance	−	**	−	−	−	.	=	*	−	*	*	**	**	**
Gasoline resistance	−	−	−	=	**	.	**	*	*	*	**	.	**	**

* Fairly good
** Good
= Fair
− Poor

Notes: 1. Attacked by water at temperatures above about 50 °C.
2. Up to 70 °C.
3. Up to 100 °C.

on the other constituents of the rubber and the method of manufacture. For this reason it is not possible to give absolute recommendations on fluid compatibility and temperature ranges etc., but *Table 4.2* gives a general indication of the properties of some of the more common rubbers. Two plastic materials are also included since they are valuable materials by virtue of high temperature resistance or extreme chemical inertness and low friction, however special design is necessary for seals incorporating materials based on plastics.

4.3.2 Materials for mechanical seals

The sliding faces in mechanical seals must be compatible with the sealed fluid, be good bearing materials and be resistant to abrasion if the sealed fluid is contaminated. A very wide range of materials has been used but *Table 4.3* indicates some of the more usual combinations for various applications. It will be noted that graphitic

Table 4.3 FACE MATERIAL COMBINATIONS FOR MECHANICAL SEALS

	Carbon			
	Lead-Bronze	*Stainless steel*	*Stellited SS*	*Ceramic*
Water (clean)	x	x		
Water (brine; boiler feed)	x		x	
Mine water			x(f)	x(f)
Water (sandy)			x(f)	
Steam		x	x	
Gases		x(d)		
Oil (mineral)	x	x		
Refrigerants		x		
Organic solvents		x	x	x

d = double seal
f = flushed seal
g = glass-loaded ptfe
Metallised carbons may be used up to 180 °C, use plain carbons above this temperature.

carbon is favoured for most applications, however when the environment is particularly strongly oxidising, or abrasive, alternatives are preferred. For the most abrasive fluids tungsten carbide is now widely used for both faces of the seal. The inertness of ptfe makes it a valuable base material, often with glass fibre reinforcement, when very reactive fluids have to be sealed.

4.4 Face seals

This is a rotary seal incorporating a pair of hard flat faces which are pressed together axially to control the leakage. *Figure 4.2a* illustrates the main features of a face seal. As drawn, the sealed fluid is on the outside of the seal (it can be on the inside), one sealing ring is fixed

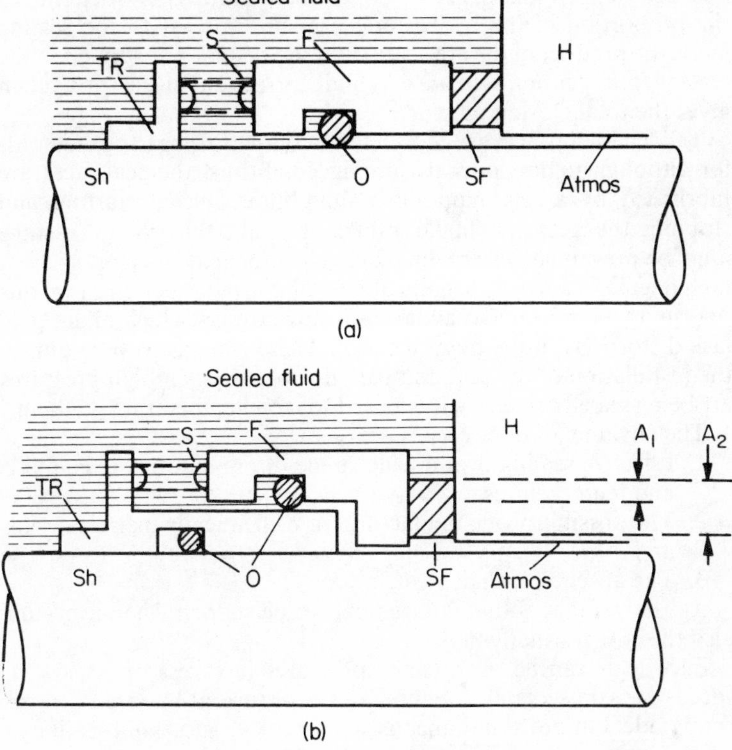

Figure 4.2. Typical face seal designs (*a*) *unbalanced;* (*b*) *balanced.*
F. Floating member
H. Housing of machine
O. Static seal (*O-ring*)
S. Spring
Sh. Shaft
SF. Stationary seal face
TR. Thrust ring

to the housing of the machine being sealed and the other is mounted on the shaft and rotates with it. The rotating member is free to move axially to some extent and is maintained in contact with the other ring by means of a spring or springs which bear against a collar on the shaft. An O-ring seal prevents leakage between the floating member and the shaft. When the system is pressurised the spring force is supplemented by hydrostatic pressure on the rear of the floating member, this prevents the sealing faces being forced apart when pressurised fluid penetrates between them.

When high pressures are to be sealed the closing force can become excessive with this design and it is usual then to offset some of the hydrostatic load by using an L-shaped floating member and a shaft sleeve or stepped shaft, *Figure 4.2b*. The area ratio, A_1/A_2, determines the proportion of the hydrostatic pressure applied to the sealing faces, the product of the area ratio and the system pressure gives the closing force per unit face area, a small correction for the spring then gives the total closing pressure.

The sealing surfaces are chosen to be compatible bearing materials for although under normal running conditions the seal faces are lubricated by a self maintained fluid film. During starting and stopping the faces may make solid contact and seizure at this stage must be prevented. The sealing faces are prepared, by lapping, to a high degree of flatness, usually about a light-band per cm. For this reason it is important to avoid imposing stresses which might give face distortions in the micron range. These can easily arise during the installation of the seal and can also occur when high pressures are being sealed or the seal is exposed to large temperature gradients.

The advantages of face seals are:

1. Effective sealing over a wide range of pressures, speeds, fluids and temperatures.
2. The possibility of manufacture from chemically inert materials and
3. The absence of shaft wear.

A disadvantage is that failure tends to be sudden when it occurs, also the cost is usually high.

Face seals can be operated at pressures in excess of 70 bars, at speeds of 30 m/sec and at temperatures of several hundred degrees centigrade, but not simultaneously. To achieve successful sealing at these extremes, special precautions have to be taken and these may include the use of double seals, coolant provision or the use of special constructional materials. Particular attention must be paid to the static seal which is often a rubber O-ring with a limited temperature range and a tendency to react with many fluids. Alternatives include the use of wedge-shaped static seals made of ptfe these may be loaded with asbestos or glass fibres for better resistance to extrusion.

For general use the seal face pair is commonly carbon-graphite and a metal such as cast iron, stainless steel or lead bronze (*Table 4.3*), but where carbon cannot be used, due to oxidation or abrasion, a hard alloy such as stellite may be used. Both ceramics and tungsten carbide are commonly used for difficult applications.

The existence of a full fluid film between the faces of a face seal has been demonstrated by various workers and it now appears that this is the usual operating mode. Under extreme operating conditions, however, this favourable hydrodynamic mode of lubrication gives way to less favourable boundary lubrication.

Measurements show that the hydrodynamic film thickness is typically 0.002 mm thick and may be as low as about 0.0005 mm.

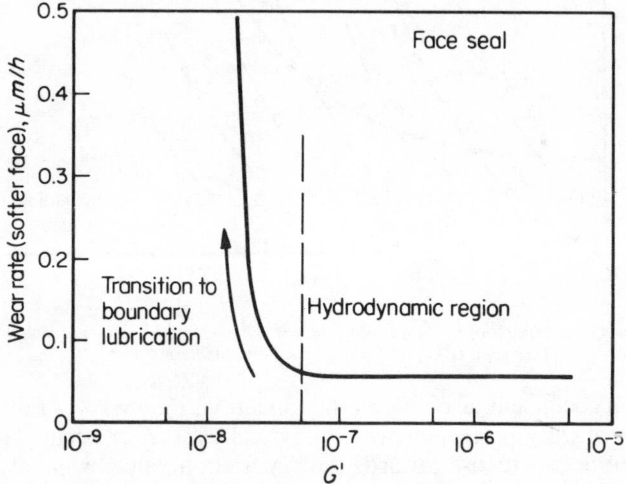

Figure 4.3. Variation of wear rate with duty parameter for a face scale (based on Summers–Smith[1])

The details of the mechanism responsible for maintaining this lubricating film are not yet fully understood but it is generally agreed that film thickness variations of the order of the film thickness must exist to account for this hydrodynamic effect. It has been found that initially flat surfaces can become wavy when run in a seal, also it has been found that significant vibration effects exist in most face seals. The surfaces can be rough as well as wavy and this might be important. Clearly such mechanisms as these could be the vital factor in maintaining the film and of the alternatives surface waviness seems the most probable.

The effect of the transition to boundary lubrication is shown very clearly in *Figure 4.3* which is based on results presented in reference 1. The wear rate can be seen to rise dramatically once the hydrodynamic film is lost at low values of the duty parameter, G'. The

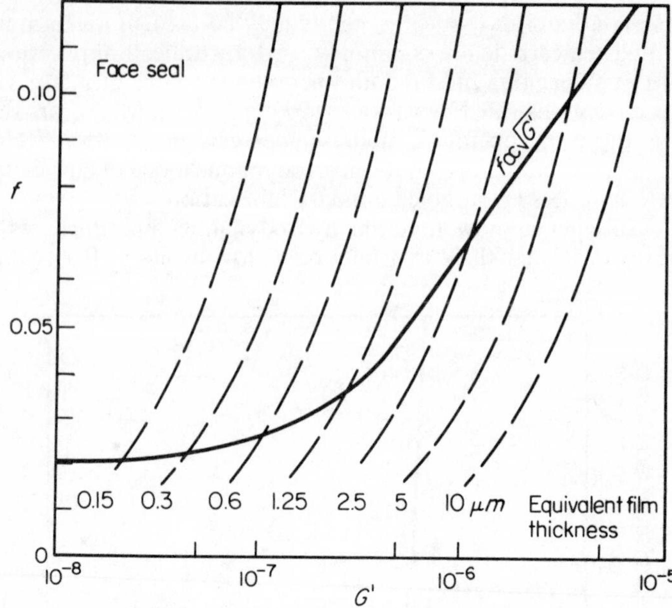

Figure 4.4. *Variation of friction coefficient with dimensionless duty parameter for a face seal,* G' $(= \mu U b \phi W)$ *(based on Summers–Smith*[1]*)*

friction coefficient also reflects the transition, *Figure 4.4*. At the low values of duty parameter where physical contact is occurring the friction begins to rise towards the dry friction value, while at high values the duty parameter and friction coefficient follow the relationship:

$$fG^{-0.5} = \text{constant}$$

When a face seal is to operate so that the interface film is close to its boiling point, special attention must be paid to the heat transfer characteristics of the seal. The film itself can be regarded as being at a uniform temperature, it is also a heat source due to the viscous shear. The thermal resistance* of the seal body depends on the structural materials and the geometry. Assuming a carbon and steel face pair, the thermal resistances of these will be, say, 40 units and 8 units. It follows that more heat flow passes through the steel member for a given temperature differential, i.e. the steel member is more effective than the carbon in removing the frictional heat. Where, as is often the case, the heat is dissipated in the sealed medium, which therefore serves as coolant, it is desirable that the

* (l_{eff}/k where k is thermal conductivity and l_{eff} is effective mean heat flow path length).

seal member with the lower thermal resistance should be the rotor since heat transfer to the coolant is more effective through the thinner boundary layer on the outside of the rotor. It is desirable to arrange the rotor design so as to minimise temperature gradients which might lead to thermal distortion of the seal face.

Sealing a fluid at high pressure leads to problems of structural deformation of the seal which can result in distortion of the seal faces

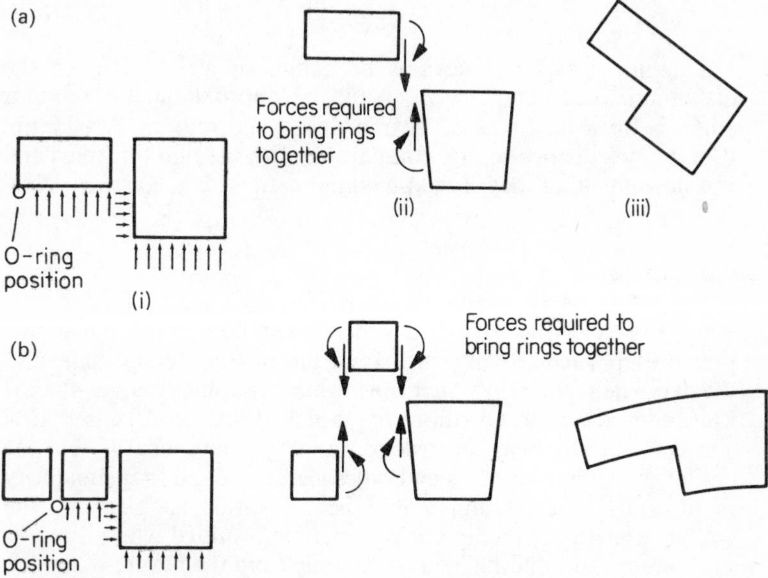

Figure 4.5. *Structural distortion of a face seal at high sealed pressures (based on Fisher)*
(i) *Floating member considered as two separate rings acted upon by sealed pressure*
(ii) *Separate rings respond to system pressure*
(iii) *Floating member deformed by pressure*

and consequent seal failure. The nature of the problem is illustrated in *Figure 4.5*. The hydrostatic forces are as shown on the left of *Figure 4.5a*, where the floating seal ring is imagined to comprise two parts, a thin walled ring and a thick walled ring. The most important effect of the pressure forces is to cause each ring to expand radially, the thin one more than the thick one. Since the rings are in fact joined, each exerts a bending moment on the other and the member as a whole is deformed so that the seal face becomes concave. When the static seal is repositioned as in *Figure 4.5b*, the bending moments are balanced and the seal face remains undistorted.

Another pressure effect exists where the seal has a face with relatively low elastic modulus e.g. carbon. The variation in the film pressure from the high pressure side to the low pressure side produces a differential degree of axial compression of the face. The simplest way of overcoming this is to reduce the axial thickness of the compressible material to a minimum, this follows from Hooke's Law, for if c is the linear compression of a member of thickness t at a pressure P, the modulus being E, then:

$$c = Pt/E$$

Thus for a carbon thickness of 12 mm subject to 70 bars, the maximum axial compression would be approximately 0.003 mm whereas for a thickness of 1.2 mm this would only be 0.0003 mm. Clearly such distortions are comparable with the film thickness and it is desirable that they should be minimised for a successful seal.

4.5 Lip seals

Rotary seals of this type are widely used for sealing oil at low pressures, particularly in gearboxes, transmissions, crankshafts and the like. The sealing lip which runs against the shaft is a vee-shaped knife edge which when run-in has a flat of 0.25 to 1.0 mm width. The details of the seal construction may vary considerably but the design shown in *Figure 4.6* embodies features found in the majority of lip seals. The circumferential helical spring supplements the contact pressure of the lip but is sometimes omitted where space is very limited, the loading then comes only from the flexure of the lip. At high speeds or for eccentric shafts the spring is essential. The metal insert shown in *Figure 4.6* stiffens the seal and is sometimes extended so as to envelop the seal and protect the delicate lip from accidental damage. Leakage round the outside of the seal is prevented by the rubber casing in contact with the housing.

Lip seal behaviour has been the subject of extensive experimental investigations[3] but the details of the mechanisms which determine the lubrication and sealing action are not yet fully understood.

In general, it is found that the coefficient of friction can be expressed in terms of the operating conditions by using a dimensionless 'duty parameter' defined as

$$\text{duty parameter, } G = \frac{\mu U b}{W}$$

where W is the force exerted by the lip and b is its effective width. Typically the contact force amounts to a pressure of about 3 bars on the lip. Under conditions of hydrodynamic lubrication, i.e. when G is large, the following relation holds approximately:

$$fG^{-1/3} = \text{constant}$$

When the operating conditions are more severe and some solid contact occurs, the friction begins to increase and ultimately reaches the value corresponding to the dry friction between lip and shaft. This is analogous to the behaviour of a face seal (*Figure 4.4*) but the power law index differs, being one-third instead of one-half. The reason for this is not known.

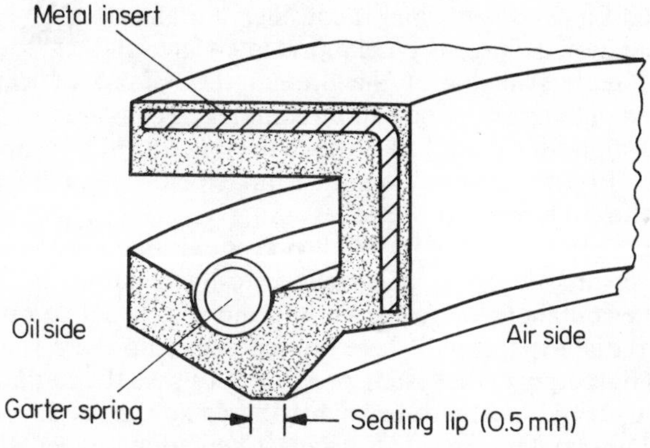

Figure 4.6. Section through a rotary lip seal

Estimates of the thickness of the lubricating film under the lip have been made from the measured friction and have also been measured directly, typically values are about 1–2 microns.

The rate of lip wear is high during the running-in period but it levels out after several hours and might then amount to about 25.10^{-6} mm^3/hr/mm. The wear rate is greater for rough shafts and it is usual therefore to have a finish of between 0.2 and 0.5 microns CLA. Dirt particles suspended in the sealed medium also accelerate wear.

The shaft itself wears under the sealing lip and it is found that the wear groove increases in depth at a rate of about 0.025 microns an hour, during the first 200 hours, the hardness of the shaft has little effect on this rate. Tests have been made with shaft finishes produced by different machining processes and, although 0.2 to 0.5 micron plunge-ground shafts are accepted practice, it is found that the lowest wear-rates and best leakage performance are obtained by vapour blasting to a finish in the 0.2 to 0.5 micron range. At smoother finishes the friction tends to increase and there is therefore no advantage in too good a finish. When run-in a shaft may be as smooth as 0.05 micron CLA under the lip, the lip itself is normally rougher than the shaft.

4.6 Moulded seals for reciprocating shafts

The design of moulded seals is governed by the materials used, usually rubber, leather or a composite rubberised fabric material. The simplest shapes are rings of circular or square section, the former is the well-known O-ring and is considered under another heading. The basic designs have been elaborated to give the wide range of moulded seals available at the present time, some of which are illustrated in *Figure 4.7*. These more complex designs are rarely made of rubber alone due to the lack of structural rigidity at high operating pressures. For the same reason, the rounded rings require a shaped backing for high pressure work.

Initial interference with the shaft prevents leakage when the seal is not pressurised. Under pressure the resilient nature of the seal causes the contact force on the shaft to increase, axial compression causing radial expansion. Where the seal has a lip there is also the direct radial action of the system pressure adding to the contact force. It is usual for the seal lip to be slightly flared so that as the pressure rises the seal contacts the shaft progressively along its length, from the lip to the heel. Some control of the interference is possible by incorporating specially shaped backing rings or by profiling the housing so that the interference does not increase indefinitely as the pressure rises.

The surface finish of the shaft against which the seal is sliding must be good if an effective seal and long life are required, between 0.1 and 0.5 microns CLA is usually recommended. The finish of the housing bore is less important and may be five or six times the shaft roughness. To prevent extrusion of the seal the clearance between the moving parts is about $1\text{–}5 \times 10^{-3}$ mm/mm dia.

Seals of this type are usually employed at pressures between 5 and 500 bars, the sliding speeds are usually quite low, mostly between 3 and 30 m/min. It is not unusual for speed or pressure to vary on each half of the cycle and this can adversely affect the leakage if the pressure is higher on the return stroke for instance. This is a result of the two-way nature of the leakage mechanism which is discussed below. It is sometimes possible to overcome this particular difficulty by inverting the seal arrangement so that the seal is fitted to the shaft instead of the housing, or vice versa.

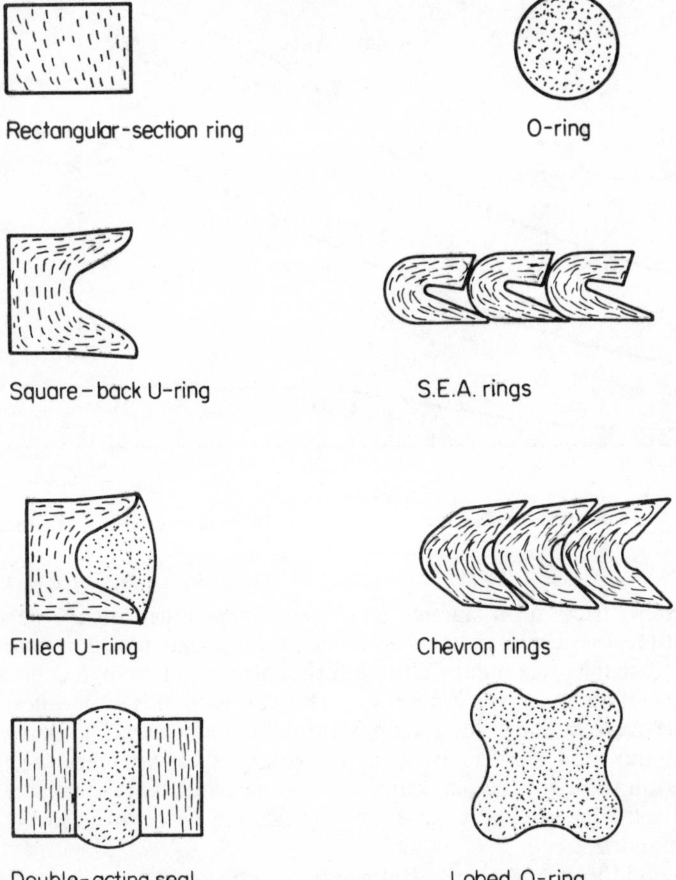

Figure 4.7. Examples of moulded seal rings for high pressure reciprocating pistons and cylinders

Another problem may arise when two moulded seals are fitted back to back on a piston. Over a period of time the seals trap a quantity of fluid between them and can build up very high pressures, this leads to extrusion or in extreme cases the retaining flange may be blown off. This problem can be overcome by venting the inter-seal space or by using a double-acting seal instead of separate ones.

Where low levels of leakage are desired, moulded rings can be stacked in series, tests indicate that there is little to be gained beyond the first two of three rings since the outer ones may then run dry and quickly fail.

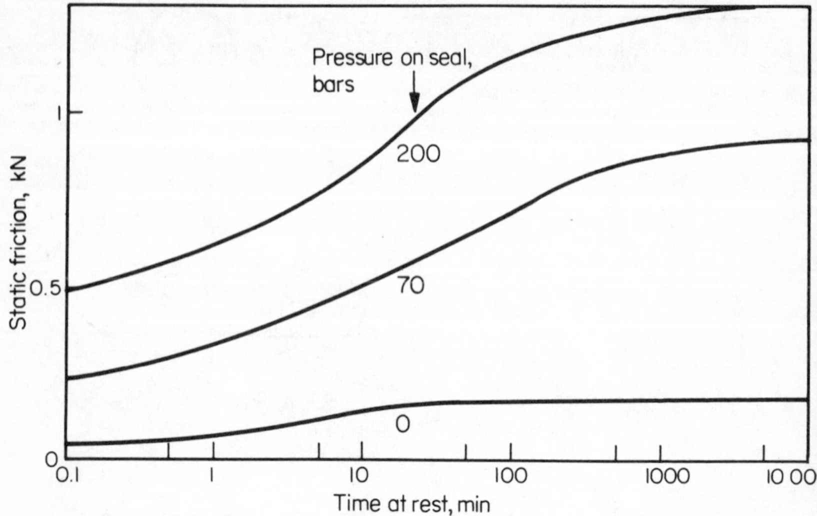

Figure 4.8. Static friction as a function of time at rest (*rubberised-fabric square section seal ring in mineral oil*)

Static friction on starting up depends very much on the period of time for which the seal has been at rest and can continue to increase for several days although the most rapid change is during the first hour or so, *Figure 4.8*. The cause of this time effect is believed to be the gradual elimination of the lubricant film between seal and shaft, the friction therefore gradually approaches the dry friction value for the particular surface materials. This effect can be a particular difficulty when a predictable level of friction is demanded.

Unlike rotary seals the leakage flow and pressure gradient for a reciprocating seal are in the same direction and this leads to quite different behaviour. As the shaft is drawn through the seal a film of the sealed liquid is drawn out and this is mostly returned to the system on the return stroke, the balance is the effective leakage.

In general, flow past the seal is due to a combination of two factors:

1. Pressure gradient flow through the lubricating film, and
2. Shear flow due to viscous drag.

These two factors are represented by the two terms on the right of the expression:

$$q = -\frac{h^3}{12\mu}\frac{\mathrm{d}p}{\mathrm{d}x} \pm \frac{Uh}{2}$$

where q is flow rate per unit length;
h is mean film thickness;
μ is fluid viscosity;
U is sliding speed; and
$\mathrm{d}p/\mathrm{d}x$ is the applied pressure gradient.

When the shaft moves outwards both terms have the same algebraic sign but on the return stroke the second term changes sign. The relative importance of the shear and pressure flow is determined by the value of the Flow Number:

$$\mathscr{F} = \frac{6\mu U X}{h^2 P_s}$$

where X is the axial length of the seal and P_s is the system pressure.

When $\mathscr{F} > 1$ the shear flow predominates and when $\mathscr{F} < 1$ pressure flow predominates. For example, with an oil of viscosity 0.07 Ns/m^2 pressurised at 70 bars, a sliding speed of 0.25 m/sec and seal of length 8 mm, the critical flow number is reached when the film thickness is approximately 10 microns. The pressure flow under these conditions is 2 ml/min, a value which is higher than is normally acceptable as leakage in practice, the implication is therefore that the film thickness is usually less than this and in consequence shear flow will predominate. Neglecting pressure flow therefore, the average volume of leakage per stroke is:

$$Q_s = (h_o - h_i)S \text{ per unit periphery}$$

where the suffices refer to the outward and inward strokes and S is the stroke length. In general the speed may differ in each direction and so the average leakage *rate* is:

$$\bar{q} = \frac{(h_o - h_i)S}{\left(\frac{1}{U_o} + \frac{1}{U_i}\right)S} = \frac{(h_o - h_i)U_o U_i}{(U_o + U_i)}$$

In practice the film thicknesses are unknown functions of the operating conditions and cannot be evaluated at the present state of the art. From experiments it is known however that for most moulded seals the leakage is given by:

$$\bar{q} = A\mu U^2$$

where A is a constant for pressures over 70 bars and the speed, U, is the same on both strokes. Observe that the leakage is worse for a thicker fluid, the viscous drag effect being greater. The friction of moulded seals is relatively insensitive to speed but tends to increase with pressure. In general it is true to say that high friction is associated with low leakage and vice versa so that it is not usually possible to work under conditions of low leakage and low friction at one and the same time.

4.7 Piston rings

Although differing from moulded seals in that they are not made from a low modulus material, piston rings nonetheless have much in common with reciprocating moulded hydraulic seals. Experiments reveal the existence of a full lubricating film during most of the cycle,

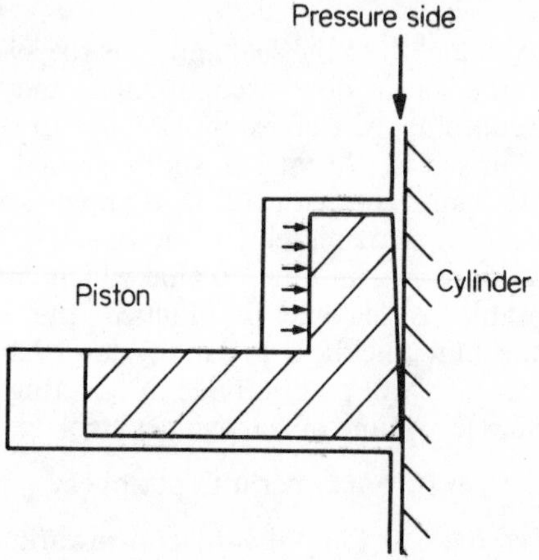

Figure 4.9. Cross-section of L-shaped piston ring which prevents inertial 'seating' at high speeds. System pressure keeps ring in contact with cylinder wall.

the film may however collapse briefly at the end of the stroke where the pressure is high. As with moulded seals, leakage is a two way process and anything which hinders the return of the extruded fluid film causes increased leakage. Thus in one series of experiments the fitting of a scraper ring with reverse taper caused the leakage to increase from 0.18 ml/hr to 490 ml/hr. If precautions are taken to avoid such adverse conditions very low leakage rates are possible, even though a large volume of oil passes the piston on each stroke.

The radial force keeping the piston ring in contact with the cylinder bore comes from two sources, the flexural elasticity of the ring, which is initially oversize for the bore, and from the pressure of the sealed system. For the latter to be effective it is necessary to ensure free access of the fluid to the back of the ring. At high speeds, inertia may cause the ring to seat against the pressure side of the groove and exclude the fluid, this may be prevented by using an L-shaped ring and groove, *Figure 4.9*.

Thermal effects can be important in piston ring seals. Thus the ring normally has a break of some 0.1 to 0.125 mm when cold but the design is such that this is almost all taken up by thermal expansion at the working temperature. In addition, the ring serves as a heat flow path from piston to cylinder in an internal combustion engine and this function may have to be kept in mind when designing a piston ring seal.

4.8 Soft packing seals

The soft packed gland is a simple and effective seal which can be used on many arduous duties, it is used on both rotary and reciprocating pumps as well as valve stems. The advantages over a

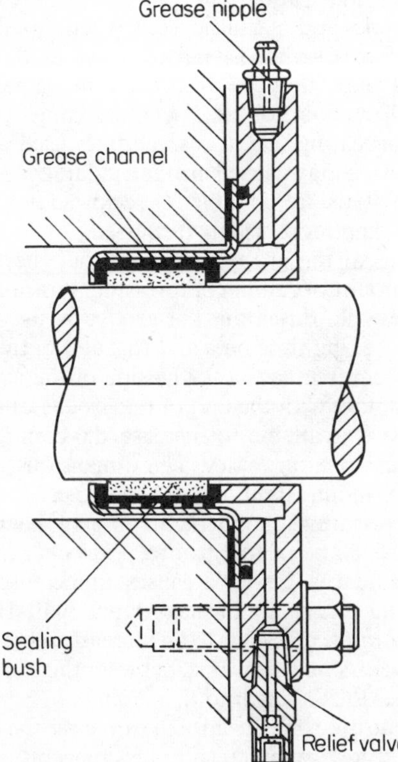

Figure 4.10. Radially-loaded soft packing seal (*courtesy, Flexibox Ltd*)

face seal are that the soft packing can be renewed without separating pump and motor units, also the period over which failure develops tends to be longer and this leaves more time for remedial action.

A wide range of packings is available but the majority are fibre-based with various fillers and lubricants, the latter may be greases, oils or such solids as graphite or molybdenum disulphide.

A development of the ordinary stuffing box which has recently become available, *Figure 4.10*, utilises radial grease loading and dispenses with the usual bolt adjustment arrangement. This gives more uniform sealing stress along the packing length and is generally more convenient.

4.9 O-ring seals

O-rings are toroidal rubber rings of circular cross-section, they constitute a simple and versatile seal when fitted in a groove, (*Figure 4.11*). The O-ring is particularly used in semi-static situations where there may be occasional movement of one part relative to another, it is not used where sliding speeds are high. Primarily a reciprocating seal, it is sometimes used as a rotary seal but excessive heating may cause permanent damage at speeds above 3 m/sec. Higher speeds are possible for reciprocating seals since the frictional heat is then more widely dispersed.

The dimensions of the groove are such that the ring is initially under a small amount of radial compression, typically 5–10% of its thickness. In the axial direction, the groove is usually about half as long again as the ring thickness and this allows the fluid pressure of the sealed system free access to the side of the ring, so that it is maintained in contact with the end of the groove and a proportion of the axial stress is available to increase the contact force on the shaft as the system pressure rises. The dimensions of O-rings and their grooves are standardised in most countries.

When fluid pressure is applied, an O-ring begins to deform, conforming to the shape of the groove, *Figure 4.12b*, and this has important consequences. Firstly it causes an increase in the contact zone between ring and shaft in accordance with Hertzian elastic theory for small strains, this leads to an increase in friction. Secondly, since there is always some clearance between the shaft and housing, the pressure forces the O-ring into this, *Figure 4.12c*, and may cause serious damage to the ring. In an extreme case the O-ring can be ejected from the groove completely. Extrusion is prevented by fitting a hard washer in the groove on the low pressure side of the O-ring,

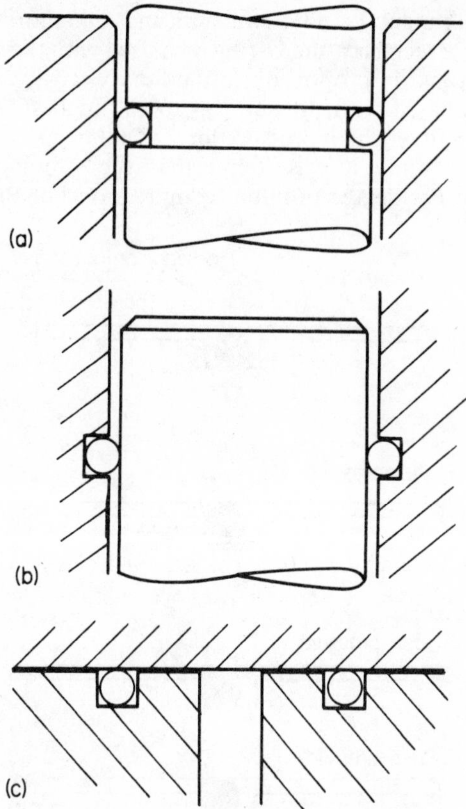

Figure 4.11. O-ring sealing arrangements
(a) O-ring as a shaft seal (b) O-ring as a cylinder seal
(c) O-ring as a flange seal (static)

selecting a material which is compatible with the shaft. In this way the effective clearance can be reduced and extrusion prevented, *Figure 4.12d*.

Since rubbers are easily deformable, yet relatively incompressible, they behave somewhat like a very viscous fluid when pressurised. For this reason the contact pressure between an O-ring and shaft is very pressure dependent, the radial stress component and the contact width together play an important part in determining the friction and leakage of an O-ring seal. Detailed analysis of this interaction is difficult and is made more so by the complex elastic properties of rubbers. The lubrication of reciprocating seals is discussed further under the heading of moulded seals (section 4.6).

In an experimental study, Iwanami and Tikamori[4] found that leakage from a reciprocating O-ring increased with the square of the sliding speed and in proportion to the fluid viscosity. The leakage rate also increased with pressure up to the highest test pressure, 100 bars, a relatively high pressure for a dynamic O-ring seal.

In another study it was found that the running friction is insensitive to the degree of initial compression but that the static

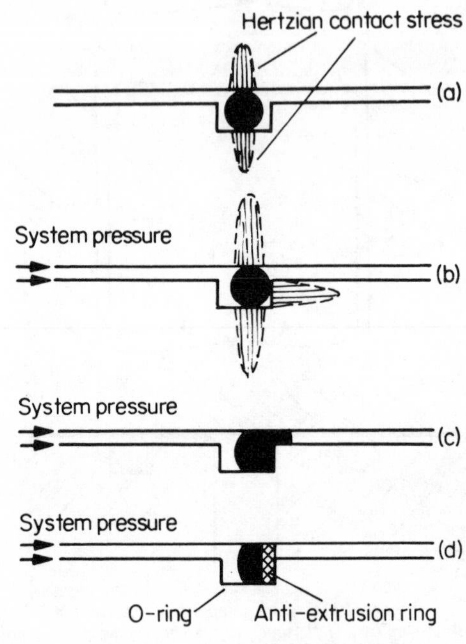

Figure 4.12. O-rings under pressure
(a) Unpressurised O-ring
(b) Pressurised O-ring
(c) Extrusion under pressure
(d) Effect of anti extrusion ring

friction on starting increases with the degree of interference. For surface roughnesses between 0.05 and 0.75 microns CLA both static and dynamic friction are less on smoother shafts. At sliding speeds greater than about 3 m/min the friction of reciprocating O-ring seals is almost independent of speed.

In recent years some variants on the basic O-ring concept have been introduced. Among these may be mentioned pressure-filled hollow metal O-rings and rings of non-circular section. Metal O-rings are employed where extreme corrosion or very high pressure or temperature problems are encountered in static applications.

4.10 Clearance seals

In a conventional seal the lubricating film between the sliding faces is controlled by relatively unknown factors and this leads to a measure of unreliability which may not be acceptable in certain applications, for instance where health hazard or plant shutdown would result from seal failure. In clearance seals this uncertainty is

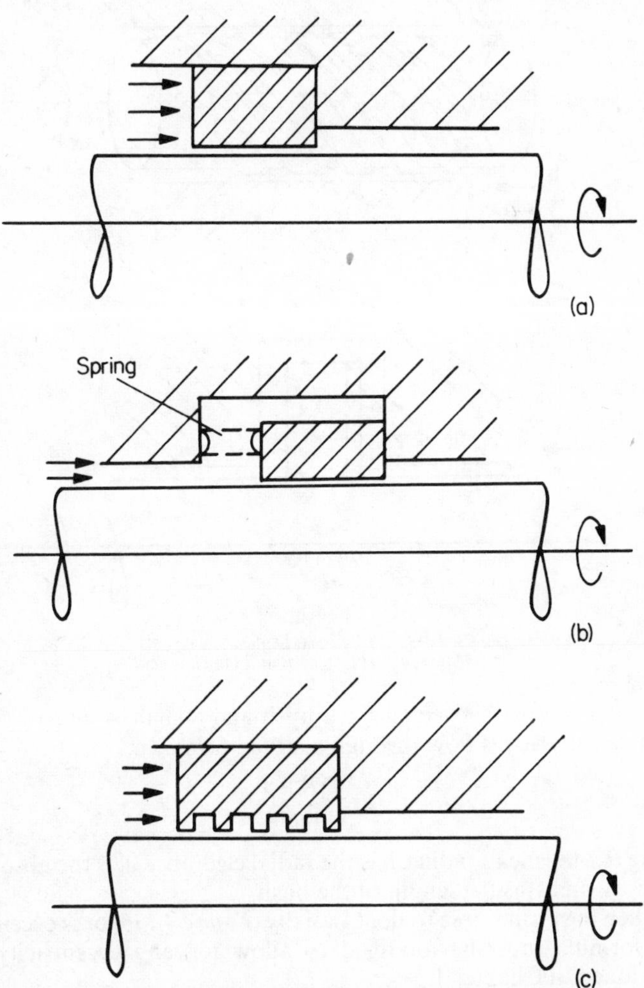

Figure 4.13. Bush type clearance seals
(a) Fixed bush (b) Floating bush (c) Labyrinth

avoided by working at larger film thicknesses, where hydrodynamic effects between the sliding parts of the seal are negligible and the clearance can be controlled by other means.

The simplest method of clearance control is to have a predetermined geometry, a close fitting bush for instance (*Figure 4.13a*).

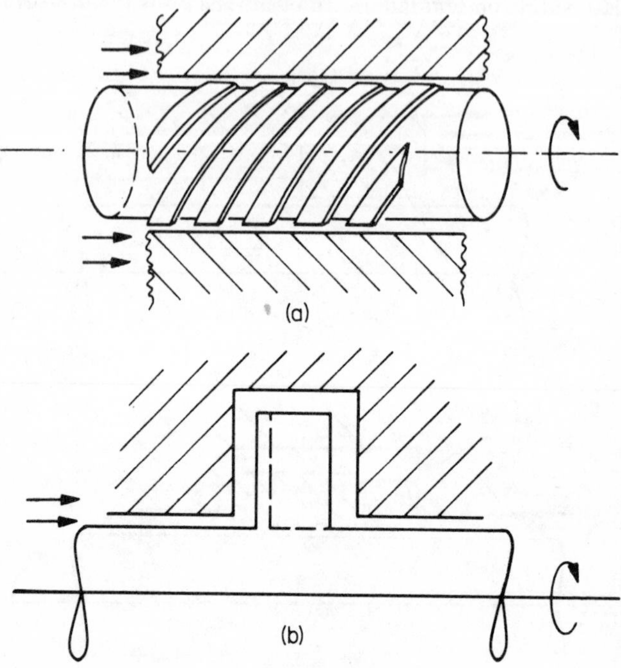

Figure 4.14. Impeller-type clearance seals
(a) Screw seal (b) Centrifugal impeller seal

The leakage flow between a fixed bush and a shaft is given by the formula for viscous flow through a straight slot:

$$Q = \frac{\pi R h^3 p}{6\mu L}$$

where R is the mean radius, h is the radial clearance, μ is the absolute viscosity and L is the length of the bush.

When the bush is free to float radially (*Figure 4.13b*) or is eccentric this formula must be modified to allow for any eccentricity as mentioned in Chapter 1.

A development of the plain bush is the labyrinth (*Figure 4.13c*), particularly used as a turbine interstage seal. In this, one or both sliding surfaces have circumferential slots in which eddies form and cause added flow resistance. Such a seal is more effective than a bush of the same length and clearance.

The disadvantage of bush and labyrinth seals is that there is always some leakage, two forms of clearance seal are however capable of low

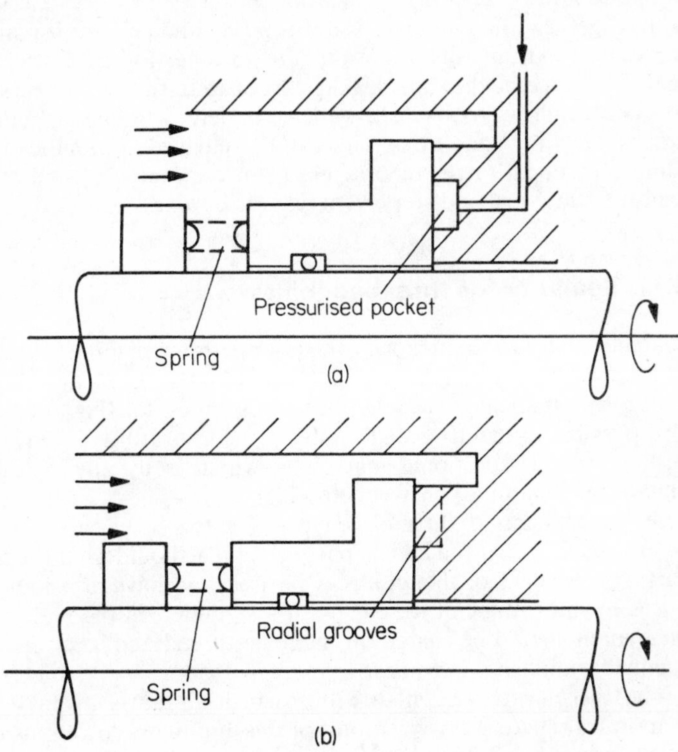

Figure 4.15. Bearing-land clearance seals
(a) Hydrostatic land seal (b) Hydrodynamic land seal

or zero leakage when rotating, both act effectively as a pump. The 'screw seal' has a helical thread on the shaft (*Figure 4.14a*) or housing bore and builds up an axial pressure gradient to oppose the system pressure. The 'centrifugal impeller seal' (*Figure 4.14b*) resembles the impeller of a centrifugal pump and produces a radial pressure gradient to oppose the system pressure.

The third type of clearance seal incorporates a coplanar bearing to control the separation of the sealing lands of what is basically a face seal (*Figure 4.15*). Both hydrostatic and hydrodynamic thrust bearings have been used. In these arrangements the working clearance is such that any hydrodynamic pressure between the actual sealing lands is small compared with the pressures in the bearing, it is therefore possible to operate at a predetermined clearance and this can be chosen to give the best compromise between leakage and power consumption.

The advantage of coplanar bearing seals is that the seal faces can come together when the shaft is not rotating and prevent leakage in the static condition, this is not possible in other forms of clearance seal. To overcome the static leakage problem in these is is necessary to incorporate an auxiliary seal which is preferably inoperative under normal working conditions. This can be achieved automatically by using centrifugal forces or pressure from the sealed system to off load the static seal and so prevent excessive wear.

4.11 Sealants for threaded joints

Sealants fall into two categories, those for permanent joints and those for joints which will have to be dismantled later. 'Cements' may be used for permanent joints and 'lutings' or 'tapes' for the latter. It is also possible to seal threaded joints by incorporating an O-ring or gasket but the additional machining which is usually necessary makes this a more expensive method.

Screw joints are of three basic types. The two screwed parts may be arranged so that the female part beds on a shoulder on the male part or vice versa, or the joint may be 'free' and have no shoulders, this is the most difficult to seal. With a shoulder on the male piece the sealant should be coated on the male thread and vice versa for a female shoulder.

The least effective cements are those which contain a solvent which evaporates. There are two reasons for this, firstly the solvent usually occupies between 20% and 60% of the original volume, so the sealant contracts excessively on setting, and secondly evaporation may be very slow in the confines of a joint and the cement is therefore slow to set. Anaerobic sealants are very quick setting and are activated when oxygen is excluded, as happens when threaded parts are screwed together. Where the joint is exposed to varying temperatures rubbery materials perform well, being able to take up the relative motion of the components of the joint. Rigid materials are liable to fracture permanently under these conditions.

When there is a possibility of exposure to chemically active substances the compatability of the sealant with these must be considered. Ptfe has desirable properties in this respect being very inert and having a wide range of working temperature.

REFERENCES

1. Summers-Smith, D., 'Laboratory investigation of the performance of a radial-face seal', *Proc. 1st Int. Conf. on Fluid Sealing*, BHRA, Cranfield (1961).
2. Fisher, M. J., 'An analysis of the deformation of the balanced ring in high-pressure radial-face seals', *Proc. 1st Int. Conf. on Fluid Sealing*, BHRA, Cranfield (1961).
3. Jagger, E. T., 'Study of the lubrication of synthetic rubber rotary shaft seals'. *Proc. Conf. on lubrication Wear, I. Mech. E.*, 409 (1957).
4. Iwanami, S. and Tikamori, N., 'Oil leakage from an O-ring packing'. *Proc. 1st Int. Conf. on Fluid Sealing*, BHRA, Cranfield (1961).

BIBLIOGRAPHY

Austin, R. M. and Nau, B. S., *The seal users' handbook*, BHRA, Cranfield, Bedford (1974).

Bisson, E. E. and Anderson, W. J., *Advanced bearing technology*, NASA, Washington (1964).

Blow, C. M., 'Elastomers for seals', in *Review and bibliography on aspects of fluid sealing*, BHRA, Cranfield, Bedford (1972).

Bowden, F. P., and Tabor, D., 'The friction and lubrication of solids'. Vol. I (1950); Vol. II (1964), Oxford University Press.

King, A. L., *Bibliography on fluid sealing*, BHRA, Cranfield (1962).

Pinkus, O. and Sernlicht, B., 'The theory of hydrodynamic lubrication', McGraw–Hill, New York (1961).

Proceedings of 1st–7th International Conferences on Fluid Sealing, BHRA, Cranfield (1961, 1964, 1967, 1969, 1971, 1973, 1975).

Chapter 5

Pumps, Motors and Transmissions

Hydraulic machines which receive energy from an external source and impart this to a fluid may be generally classified as pumps. A machine which receives an energised fluid (the energy being pressure or velocity or both) and converts this energy into some form of work such as a shaft rotation against a resisting force, may be classified as a motor.

In this chapter attention will be focussed mainly on those machines having a rotating shaft but it should be remembered that a reciprocating jack may also be classified as a pump (plunger pumps or ram pumps) or as a motor (linear motor or jack) depending on the state of the fluid entering it and the type of force applied to or taken from its shaft.

5.1 Hydrokinetic and hydrostatic pumps

Most types of pumps and fluid motors may be described by either the term Hydrokinetic or Hydrostatic machinery. The former are seldom encountered in fluid power systems and only a brief description of their operation will be given. Further information on this type of machine is available in several texts, e.g. Addison[1] and Rouse and Ince.[2]

5.1.1 Hydrokinetic pumps

A simplified diagram of a common form hydrokinetic or roto-dynamic pump is shown in *Figure 5.1a* and will be recognised as a centrifugal pump.

The shaft and rotor, which is usually built up from radial vanes, are rotated, thereby imparting energy to the fluid in the vicinity of the vanes. The energy transferred to the fluid results in an increase in the local fluid velocity and if the resulting flow of the fluid is restricted then its pressure rises. It should be noted that the discharge may be completely blocked off without causing any immediate physical damage to the unit and this is an outstanding characteristic of hydrokinetic pumps. Naturally in small, high speed units this action results in heating up and subsequent failure but generally the fact

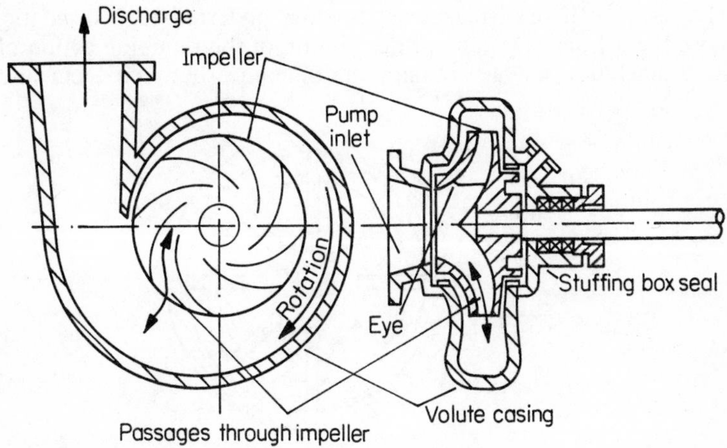

Figure 5.1(a). Centrifugal pump

that the machine may run with the delivery shut off is an attractive feature since it permits rapid reductions in the discharge to be made and it is unnecessary to wait for the shaft to stop rotating before closing the discharge valve.

The pressure or discharge head is proportional to the square of the shaft speed and in general it is limited to twenty or thirty bars. Modern designs of boiler feed pumps are made for higher pressures and units may be connected in series to raise the final discharge head. Compared with hydrostatic or positive displacement designs the overall efficiencies are generally low, seldom rising much above 70 to 80%.

Centrifugal pumps are used when either very large flow rates at low or medium output pressures are required or when the flow rate is small and the power loss due to the inefficient action is not important. They are also popular for use with contaminated fluid since a limited amount of wear of either the casing or vanes does not have as large an effect on their performance as it would on that of a positive displacement machine. In addition, their characteristics are little affected by limited changes in fluid viscosity but in common with all types of pumps they are susceptible to the effect of inlet restrictions and will generally need priming if the level from which they are pumping is too far below their casings.

5.1.2 Positive displacement of hydrostatic pumps

The term positive displacement is often preferred to hydrostatic, since the former gives an apt description of the pumping action of such machines and accentuates the difference between them and

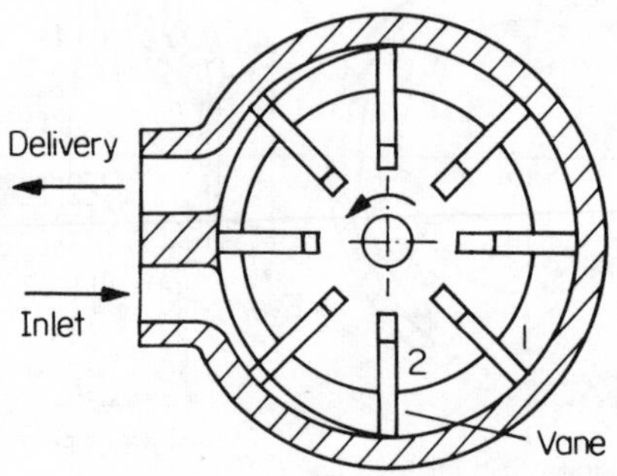

Figure 5.1(b). Vane pump

hydrokinetic machines. A typical design of positive displacement pump is shown by the simplified diagram in *Figure 5.1b*, where it can be seen that the volume received by the pump and bounded by the two vanes 1 and 2 will eventually be discharged as they pass the delivery port.

Since most liquids are virtually incompressible there is no need to reduce the volume of the fluid contained between two vanes until it has access to the discharge port and the increase in pressure is achieved by the presence of a restriction downstream of the pump outlet. It is very important to note that outlet must not be closed off until the rotor has stopped moving since this is the only path that the discharged liquid can take and, unlike hydrokinetic machines, the unit would be severely damaged if this were done.

Internal leakage has an important effect on the action of such machines and together with unbalanced, hydrostatic forces it generally limits their maximum power. However, very high operating pressures can be obtained and although vane pumps are often limited to 150 bars, other types of positive displacement pumps can produce output pressures of up to 800 bars. Flow rate, initially, varies directly with speed but as the output pressure rises, the internal leakage from the high pressure side to the low pressure side increases and the delivery gradually falls off. The head/flow characteristics for a constant input speed of both a centrifugal and a vane pump are shown in *Figure 5.2*, but it is again emphasised that it is undesirable

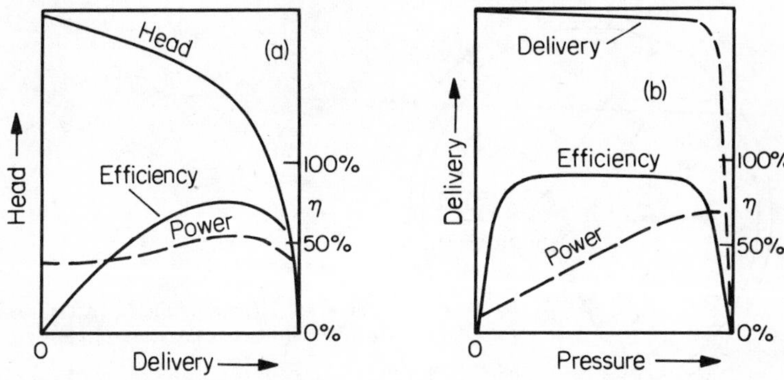

Figure 5.2. Constant input speed pump characteristics (left) *Centrifugal;* (right) *Vane*

to attempt to obtain that portion of the vane pump's characteristic shown by the dotted line.

Positive displacement pumps are generally used when low to medium flow rates of, say, 2 to 400 litre/min are required at medium to high or very high pressures of, say, 20 to 200 bars or even 800 bars.

5.2 Positive displacement machine designs

Except for occasional low-pressure auxiliary systems, hydrokinetic machines are not used in fluid power engineering systems. For this reason the emphasis, in the following paragraphs, is on the positive displacement designs which are generally encountered in fluid power circuits.

5.2.1 Early designs

One of the earliest references to force intensification is to be found in the writings of Leonardo da Vinci who described exactly the principle of the hydraulic press. Although the development of this device is generally attributed to Bramah, who was one of the earliest hydraulic development engineers, it should be remembered that the basic principles of some of his inventions were described, often in great detail, by earlier workers and as a result his claim is to the more valuable ability to develop ideas rather than to inventiveness.

Variable delivery, positive displacement machinery appeared in the latter half of the 19th century[3,4] and among the earliest is the Rigg motor or 'engine'.[5] The design, shown in *Figure 5.3*, was for

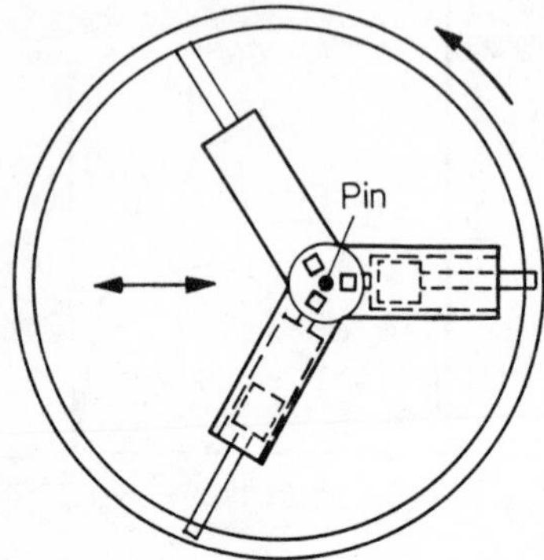

Figure 5.3. The Rigg Engine (pin moves to adjust stroke and does not rotate)

driving capstans and other machine tools using water and was intended to be economical at light loads. Motor designs of this nature became obsolete following the development of the electric motor and it was not until there was a desire to use infinitely variable gears in motor cars in the closing years of the 19th century that further progress was made. Designs of interest are the Hall gear (1896) and the Manly gear (1905), both of which had radial pistons and individual valves for each cylinder. The former design had a rotating cylinder block and shaft, whilst the latter had a rotating crank and fixed cylinder block.

Later designs in the UK, included the Janney–Williams gear (1907), the Hele–Shaw gear (1908) and the Pittler gear (1907) and Lentz gear in Germany.[6] It was when these gears for automotive applications first appeared that oils rather than water were first used as the driving fluid.

Most of these early pump and motor designs were evolved from the ancient ram pump in which a piston reciprocated in a cylinder and in general they may be classified as either the *radial piston* or the *axial piston* design depending on the position of the axes of the pistons relative to the axis of rotation of either the cylinder block or driving shaft.

5.2.2 Later designs

More recent designs include the gear and vane pumps. Examples of their operation together with those of similar designs, described by both Beacham[3] and Anderson[6] are shown in *Figure 5.4*. Both designs represent attempts to reduce the production cost of machines

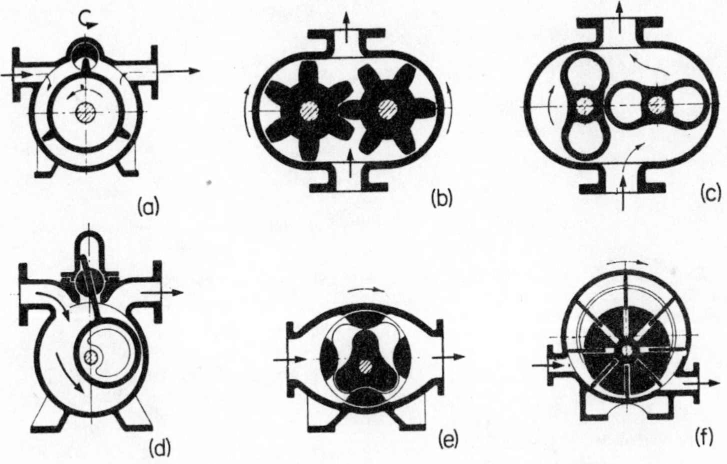

Figure 5.4. Examples of some positive displacement machines
(a) rotary abutment *(b) gear* *(c) lobe* *(d) rocking slider* *(e) multi-lobe* *(f) vane*

but at the same time retaining the advantages of positive displacement. The in-line piston pump shown in *Figure 5.5*, which is produced today, is generally for special, high-pressure duties where the attraction of the small number of moving parts compensates for the lack of symmetry and the consequently heavy bearing loads. In effect it is probably closest to the original ram pump design.

It is perhaps remarkable how little material is published on the subject of the design of such machines and since with very few exceptions there is little to guide the student in this art, one can only conclude that competition coupled with a lack of detailed knowledge are the main reasons for this situation. Noteworthy exceptions are references 7 and 8. The former classifies machines according to the type of valves or ports used and, although this method has certain attractions, the method used by Shute et al[9] is perhaps more easily understood and a similar method will therefore be used here.

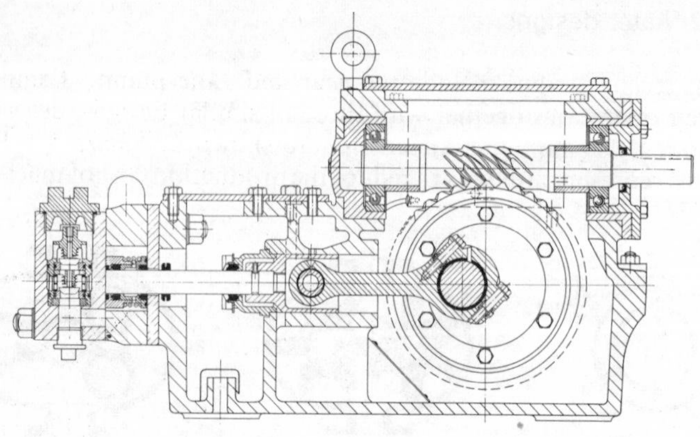

Figure 5.5. 200 bar plunger pump for mining applications (Bonser Tristram; U.K.)

Table 5.1 TYPES OF PUMP AND MOTOR DESIGNS

Type of design	*Pumps*		*Motors*	
	Variable	*Fixed*	*Variable*	*Fixed*
1. *Axial Piston*				
(a1) Rotating cylinder block	√	√	√	√
(a2) Rotating swashplate	√	√	√	√
(a3) Tilting head	√	√	√	√
2. *Radial Piston*				
(b1) Internal cam	√	√	—	√
(b2) External cam	√	√	—	√
3. *In line Piston*	√	√	—	—
4. *Vane*				
(c1) Rotary vane	√	√	—	√
(c2) Inverted vane	—	√	—	√
5. *Gear*				
(d1) External gear	—	√	—	√
(d2) Internal gear	—	√	—	√
(d3) Rotary abutment	—	√	√	√
(d4) Screw	—	√	—	√

Positive displacement pumps and motors may be divided into 'variable' and 'fixed displacement' type and generally each of these designs may be subdivided into the groups as shown in *Table 5.1*. It should be noted, however, that there are very few variable displacement vane-type designs and even fewer variable delivery gear-type ones particularly for use as motors. The method of classification used in *Table 5.1*, is based on current British designs but it is almost certain that where the absence of designs (as denoted by the dash) is indicated there are probably some machines which fall into these categories. Diagrams showing the principle of operation of some of these designs are given in *Figure 5.6* but it is emphasised that this classification is by no means exhaustive and additional information is available in references 7 to 13.

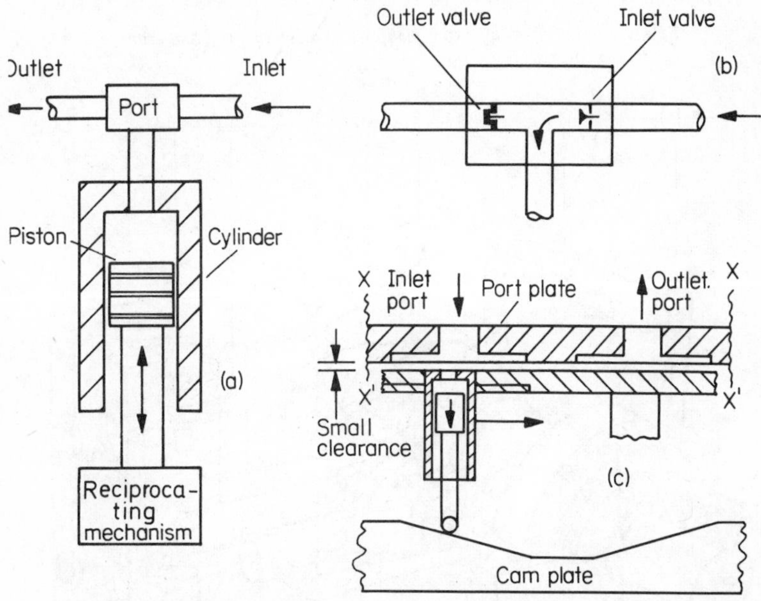

Figure 5.6 (a) to (l). Development and further examples of positive displacement machines

(a) Basic components of piston pump
(b) Poppet valve arrangement
(c) Port plate system (section of unwrapped circular plan machine)

The basic components of any piston pump are shown in *Figure 5.6a* and comprise a piston and cylinder unit, a valve or porting arrangement and a reciprocating mechanism. Two common forms of valving arrangements are shown in *Figures 5.6b* and *c*, the former being popular in large, slow speed machines and the latter, which represents the 'unwrapped' section of a multi-piston machine having a circular plan view, being common in smaller, high speed units. Some machines naturally have the poppet-type valves, shown in *Figure 5.6b*, in the actual cylinder head thereby reducing the trapped or dead volume of fluid to a minimum (i.e. reducing compression

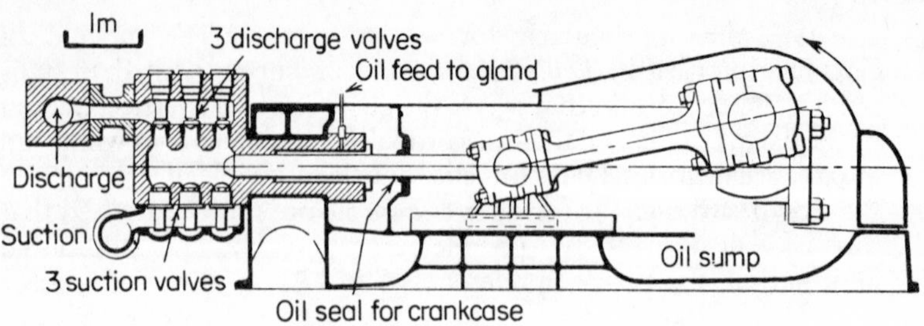

Figure 5.6(d). Piston or ram pump with connecting rod and crank

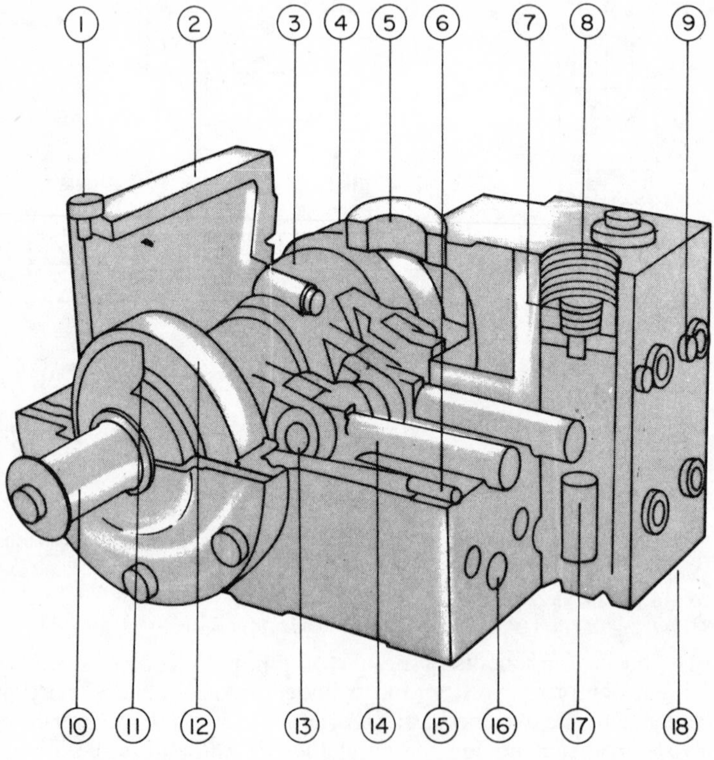

Figure 5.6(e). Eccentric-driven triple in-line piston pump

1. *Casing bleed*
2. *Casing*
3. *Connecting rod*
4. *Bearing housing*
5. *Drain*
6. *Spigot*
7. *Delivery valve*
8. *Valve seat*
9. *Primer*
10. *Shaft*
11. *Seal*
12. *Bearing*
13. *Crankpin*
14. *Piston*
15. *Cylinder block*
16. *Inlet port*
17. *Inlet valve*
18. *Valve block*

(*Cury, France*)

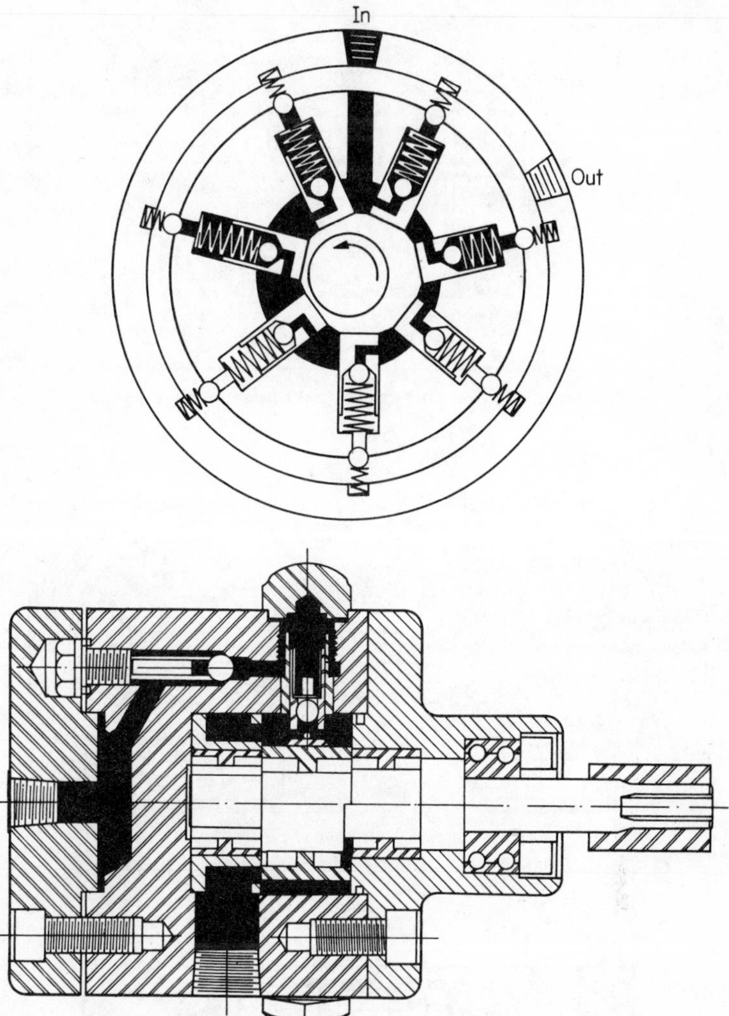

Figure 5.6(f). Constant delivery radial piston pump (Racine, U.S.A.)

losses). Unless a piston 'spider' return plate were present or a boost pressure were applied to the inlet line of the unit shown in *Figure 5.6c*, a return spring would probably be present between the piston crown and cylinder head or between the bottom of the cylinder skirt and a ring around the piston rod.

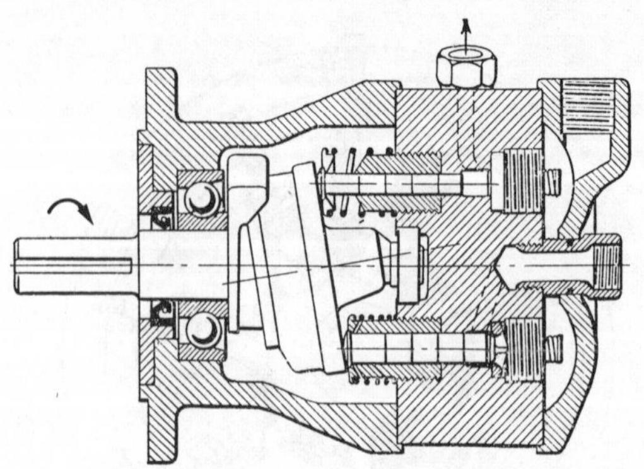

Figure 5.6(g). Constant delivery fixed angle swashplate pump

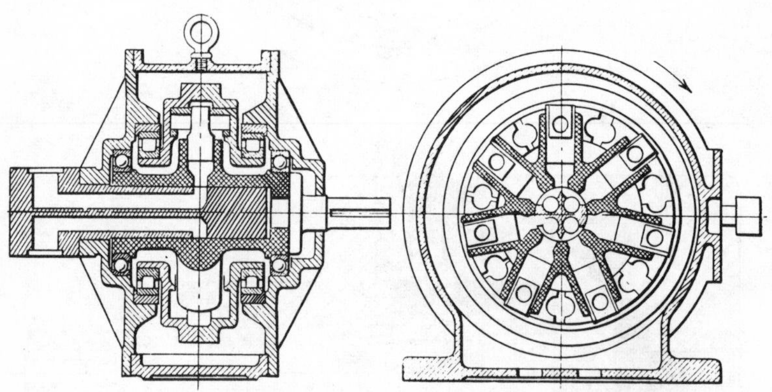

Figure 5.6(h). Variable delivery radial piston pump

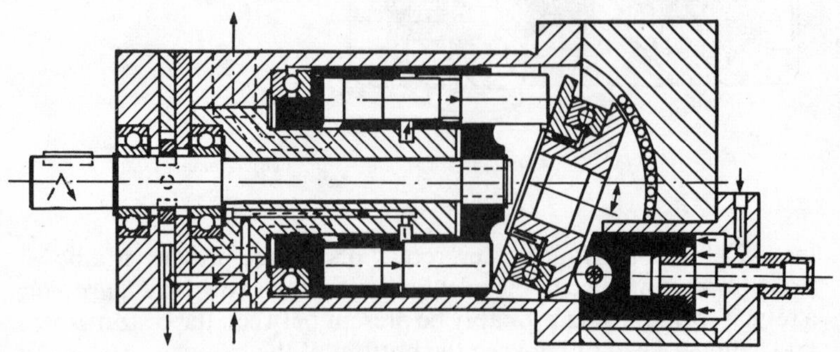

Figure 5.6(i). Variable delivery axial piston pump (Heller, Germany)

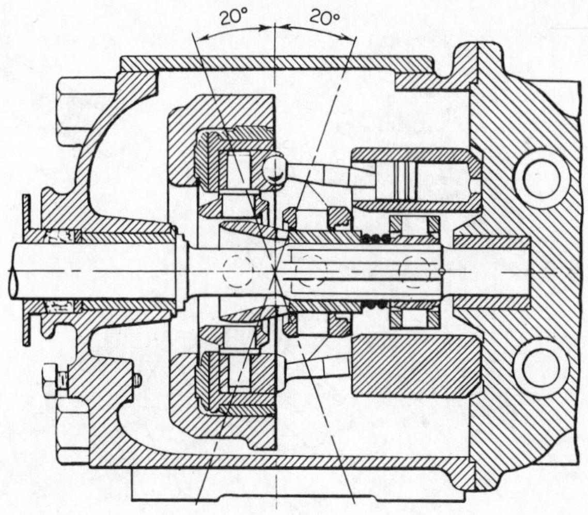

Figure 5.6(j). *Variable delivery axial piston pump (V.S.C. Co. U.K.)*

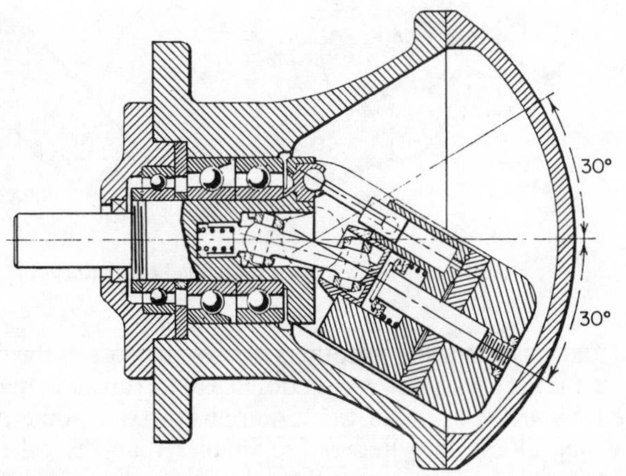

Figure 5.6(k). *Tilting head variable delivery axial piston pump (Thoma)*

Common types of reciprocating mechanisms are the piston rod and crank (see *Figure 5.6d*); a series of cams (or eccentrics) or a single eccentric on a shaft with the pistons arranged either in line in the case of the first two devices (*Figure 5.6e*) or radially with the latter (see *Figure 5.6f*) or the fixed angle swash or cam plate as shown in *Figure 5.6g*. None of these designs has a variable stroke and hence the delivery is constant unless the speed of the shaft can

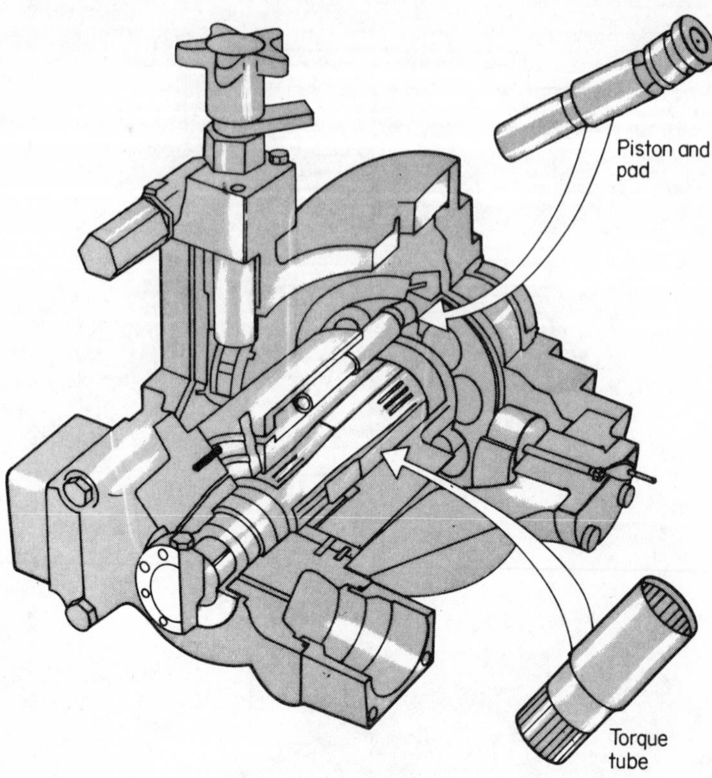

Figure 5.6(l). Variable delivery axial piston slipper pad pump (*Racine, U.S.A.*)

be varied and this naturally usually increases the cost of the driving unit. Since it is very wasteful (see section 5.5.3) to run such machines at full delivery irrespective of the required flow rate some designs exist in which alternate cylinders are smaller than the others and are ported separately. As a result the delivery can be varied in three steps from q_1 to 1 to $(1+q_1)$ where q_1 is the ratio of the swept volume of the small cylinders to that of the larger ones. This is readily achieved by lifting the inlet valves of either set of cylinders. (Further details of the economic aspects of the fixed and variable delivery machines are shown in *Figure 5.9*, page 119.)

Detailed descriptions of pumps suitable for working at high pressures and delivering large flow rates have been given by Roberts.[12] For high power applications (i.e., up to several thousand h.p.) he suggests that it is usual to employ a three-throw design, such as that shown in *Figure 5.6d*, rather than a five-throw machine for although the flow rate ripple from the former is more than three times greater than that from the latter the comparative simplicity of its design more than outweighs this disadvantage. Although few fundamental changes have occurred in recent years in the design of such pumps full advantage has been taken of modern trends in the design and operation of several of its components. For the high pressure water pump shown in *Figure 5.6d*, this includes:

(a) A high ram speed (about 200 m/min).
(b) Enclosed crankcase with forced lubrication for both crankshaft and connecting rod bearings.
(c) Lantern glands with metered oil feed for each ram.
(d) Integral valve and cylinder block forged from mild steel.
(e) Each cylinder has multi-valving and the valves are shaped to give low flow losses.
(f) Valves and their seats are made from hardened, stainless steel.
(g) Drive provided by a 2 000 kW A.C. electrical motor via a single reduction, double helical gear.
(h) Flywheels enabling a peak output of 6500 h.p. to be reached.
(i) Air pressurised reservoir with a total system volume of 60 000 litres.
(j) The water contains about 1% soluble oil.

The output of the pump shown in *Figure 5.6d* is 6000 litre/min at 150 bars and is used for the direct operation of a 7000 ton forging press. The high cost of this type of machine is offset by its low maintenance demands (e.g., the ram packings require replacement only once per year and the valve surfaces do not require attention even after ten years operation).

Significant contributions to the design and development of numerous fluid power mechanisms were made in the United Kingdom during the first half of this century by Dr. H. S. Hele-Shaw and his colleague T. E. Beacham. It was only natural that they should turn their attention to the design of variable, positive displacement pumps and motors, one of which is shown in *Figure 5.6h*. In his description of such pumps, which he designated as 'piston-type with non-rotating crank (for oil only)'[13] Beacham stated:

> 'This is the most important type. As the crank is fixed the cylinders must rotate so that some form of *rotary valve* is the most convenient and cylindrical or face valves are usually adopted. *Speeds* are limited by centrifugal force on the rotating pistons but are high enough to permit of direct drive from electric motors.

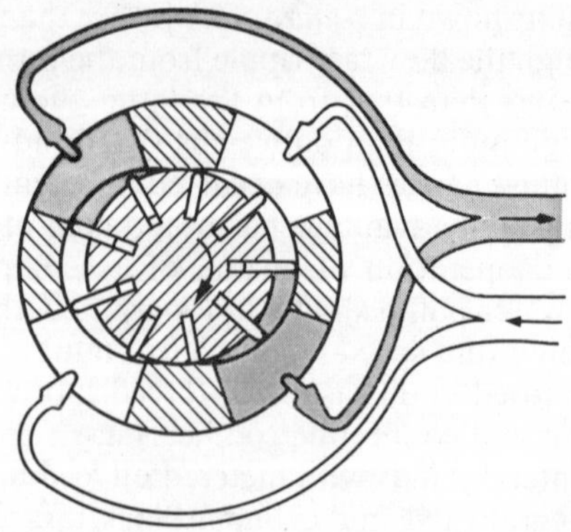

Figure 5.7 (a) to (c). Positive displacement machines, not utilising pistons, showing reservoir arrangement
(a) Principle of balanced vane type machines

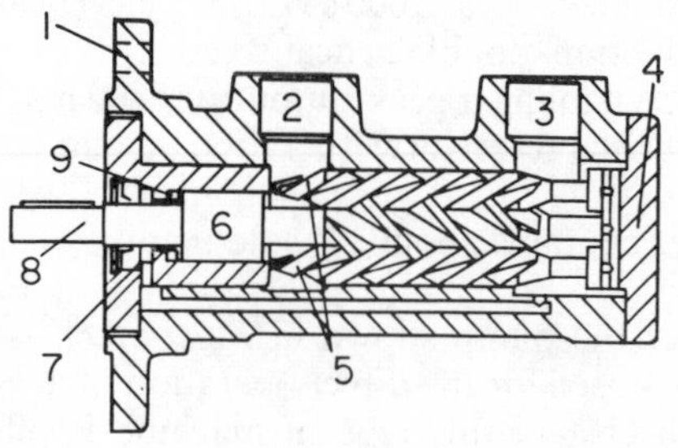

Figure 5.7(b). Typical screw pump

1. *Mounting flange*
2. *Outlet port*
3. *Inlet port*
4. *End cover*
5. *Side rotors*
6. *Bearing*
7. *Shaft seal*
8. *Centre rotor*
9. *Thrust washer*

(SIG, Switzerland)

Pumps of this type are made in sizes up to 400 h.p. In general, as compared with the fixed delivery types, *efficiencies* are somewhat lower and the noise of operation greater.'

Later designs, which are now generally of the axial piston variety, are shown in *Figures 5.6i* and *5.6j*, both machines having rotating cylinder blocks and the latter having two universal joints. In addition the tilting head or Thoma pump is shown in *Figure 5.6k*, and the more recent 'slipper-pad' pump is shown in *Figure 5.6l.*

There are now numerous other designs of positive displacement pumps available and *Figure 5.7a* shows how the radial thrust on the rotor of a vane pump may be balanced whilst *Figure 5.7b* shows the basic components of the balanced, triple-rotor, screw pump, the

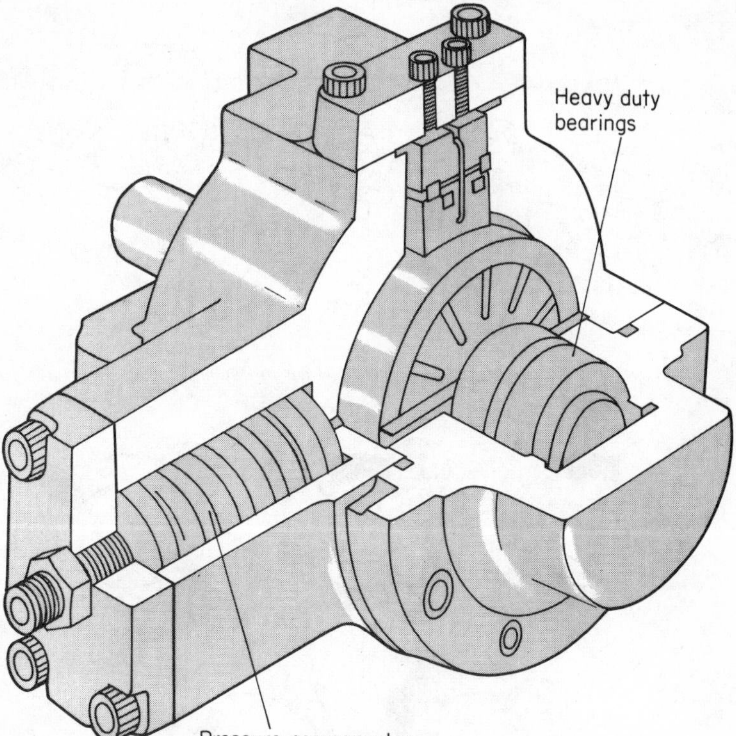

Heavy duty bearings

Pressure compensator

Figure 5.7(c). Variable delivery vane pump (*Racine, U.S.A.*)

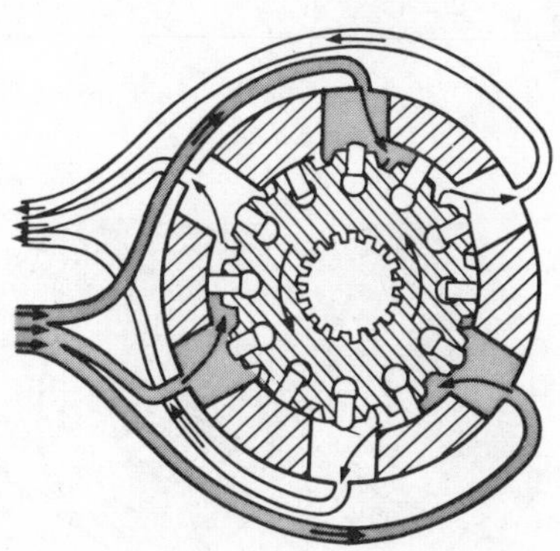

Figure 5.7(d). Balanced vane motor (*Alfred Teves, Germany*)

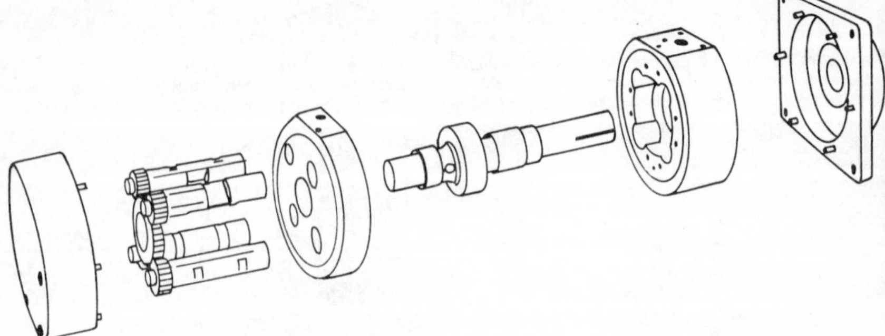

Figure 5.7(e). 'Rol-Vane' hydraulic motor (Hartman Manufacturing Co., U.S.A.)

Figure 5.7(f). Principle of the 'Rol-Vane' hydraulic motor

Figure 5.7(g). Multi-lobe pump (Marrel-Hydro, France and Double-A Products, U.S.A.)

use of which, as with all screw-type machines, gives a rippleless delivery rate.

Neither gear nor vane pumps are usually available as variable displacement machines but exceptions do exist as is shown by the variable delivery vane pump illustrated in *Figure 5.7c*. Two novel

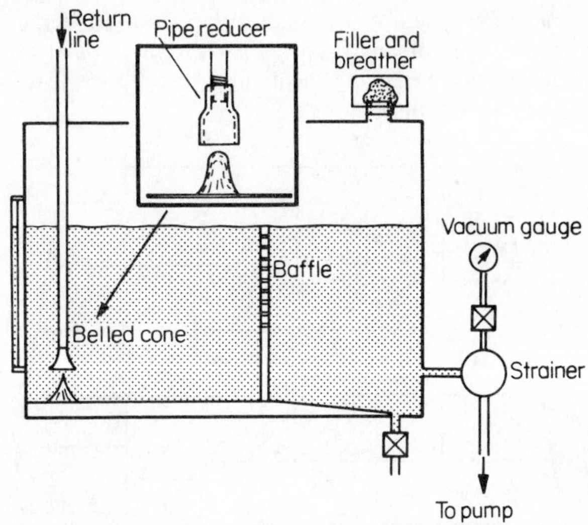

Figure 5.7(h). Suitable reservoir for pumps

forms of motor are shown in *Figures 5.7d* and *5.7e*, the latter being made under licence in the U.K. A multi-lobe machine manufactured both in France and the USA is shown in *Figure 5.7f* and radial-type machines have been made occasionally which have balls instead of pistons. However, the line contact between the ball equator and the cylinder wall generally gives rise to poor sealing characteristics and the small stroke, which is limited to less than half the ball diameter, have so far prevented this machine from gaining popularity.

Before concluding this section on various types of machines it is perhaps appropriate to include a diagram of a suitable reservoir arrangement since lack of inlet head can give rise to serious problems with any type of pump.

The reservoir system shown in *Figure 5.7h* and suggested in Reference 14 is perhaps the best since it overcomes many, if not all, of the problems encountered with the inlet conditions and with little thought the nature and purpose of each component is readily apparent. Further details of another suitable form of reservoir are given in *Figure 2* of Reference 15.

5.2.3 Other design features

Numerous variations in each type of design exist and a review of some of the more novel features of these is given by Hadekel[7] and by Cooper.[16] One noteworthy idea concerns axial piston machines in which it is possible to utilise the pistons as spool valves such that each one meters the flow from another piston 90° further round the swashplate. This arrangement is shown in *Figure 5.8a*, together with a method of balancing the axial thrust along the drive shaft by having two sets of pistons.

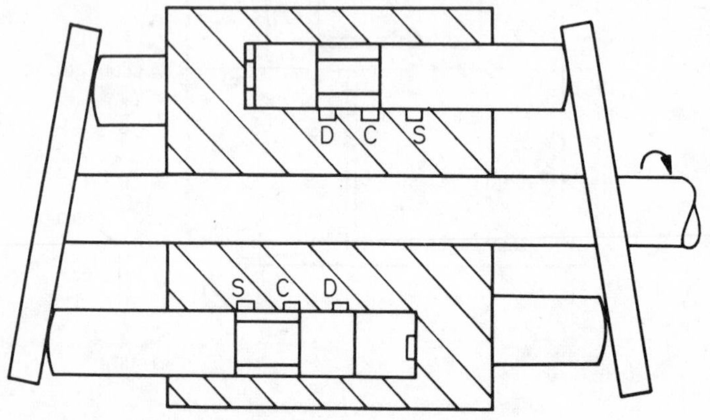

Figure 5.8(a). Pump utilising its pistons as valves (D. Delivery port: C. To cylinder of next piston; S. Suction port)

A particularly novel type of radial piston motor is described in Reference 17 in which the profile of the surface actuating the pistons has been derived mathematically such that the flow rate to and from the machine is constant for all angular positions of the shaft and, as a result, the motor produces a constant output torque. Another interesting design is described in References 17 and 18 in which an epicyclic gear device is utilised to provide a high flow rate/weight ratio.

One of the most vital parts of many axial piston machines is the valve plate where continual sliding motion must take place in a region subjected to the supply and the delivery pressure. As a result the leakage past the valve plate lands must be strictly controlled but if the clearance between the adjacent faces of the cylinder block and valve plate (*Figure 5.6c*) is too small then the viscous drag losses will become very large since they vary with the reciprocal of this clearance. A compromise is therefore sought in which the sum of the leakage loss (proportional to the cube of the clearance) and the viscous drag loss (inversely proportional to the clearance) is a minimum and this condition is examined in more detail in section 5.7.4.

In the Sunstrand design the relative velocity between the faces of moving parts is reduced by using a floating valve plate as shown in *Figure 5.8b*. The valve acts as a form of D-type sliding valve and

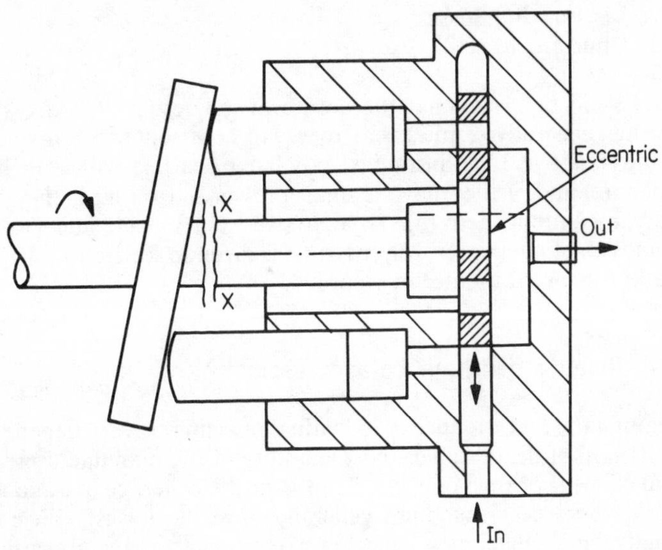

Figure 5.8(b). Floating valve plate design. (Note that the shaft has been rotated through 90° at XX to show TDC and BDC position of pistons) (Sunstrand, U.S.A.)

is free to rotate on the eccentric. Only the frictional force between the valve plate and the eccentric tends to produce rotation of the former and this results in it having a much lower rotational speed than that of the shaft. Its viscous drag losses are therefore reduced and smaller clearances become practicable, although it should be noted that both leakage and drag will occur on both sides of the valve plate.

Many other types of designs are described in Reference 8 but, in general, most commercially produced machines fall into one of the groups under the five main headings given in the previous section.

Much of the recent research and development activity in this field has naturally been devoted to improving the efficiency of the machines and, as a result, the various types of power losses, which occur in such machines, have received a considerable amount of attention. These can generally be classified as:

(a) Bearing losses;
    Ball races;
    Hydrostatic bearings;
    Hydrodynamic bearings.
(b) Leakage losses,
(c) Viscous drag losses,
(d) Churning losses.

It should be noted that the nature of the design of hydrostatic bearings generally implies that they will contribute to both items (b) and (c) above. In addition hydrodynamic bearings will contribute to item (c) and, if they have a small pressurised oil feed, they will also contribute to item (b). An analysis of the leakage and viscous drag losses of an axial piston machine is given in References 18 and 19 and is referred to later in section 5.7.

5.2.4 Relative performance and cost

Although the performance of individual machines is very dependent on the skill of the designer and the ability of the manufacturer, the usual efficiency range of each group is in the order, gear, vane and piston where gear machines generally have the lowest efficiency. Usually, small machines have lower efficiencies than larger units which follows from the fact that some of the losses are proportional to a linear dimension or an area, whereas the power of unit varies with its volume or the cube of a linear dimension. This argument holds provided it is assumed that the working clearances remain approximately constant when the size is increased.

The use of pressurised side plates in gear pumps[20] can result in overall efficiencies as high as 88% but usually a figure of 70–80% is obtained and this is sometimes acceptable since it is offset by the lower cost of such machines.

The value of the volumetric efficiency of vane pumps varies considerably from one design to another and for standard designs it can generally be anywhere between 70 and 90% at 70 bars. A figure of, say, 85% represents the probable average value. High pressure, single stage units can give results greater than 90% with an overall efficiency of 80% at a pressure of 180 bars. The overall efficiency of two stage machines can be anywhere between 55–85%.

High volumetric and overall efficiencies are a particular feature of piston pumps and values of 98% have been claimed for some axial piston type units. More generally encountered values are in the low to middle nineties but the higher production cost tends to offset this attraction.

The overall efficiencies of motors tend to be slightly lower than those of the corresponding pump designs though this is not always

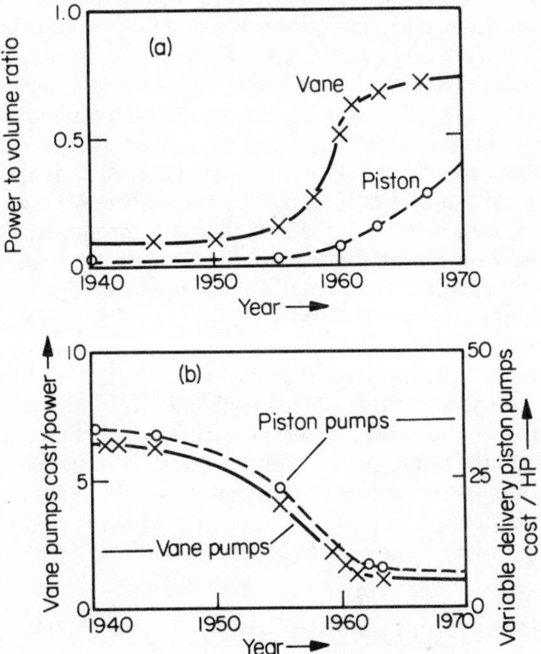

Figure 5.9. Effect of design and production improvements on (a) Power/volume ratio (b) Cost/power ratio

the case. The attractions of a high power to weight ratio and a low production cost make gear motors popular but the accompanying relatively high starting torque, relatively low efficiency, poor reversal under load characteristic and low maximum speeds and powers should be noted.

Provided a method of preventing excessive shock loads is used, it is possible to stop vane motors suddenly and even to reverse them under load. Their power output is generally limited to 20 or 30 h.p., but use of the inverted vane design enables values as high as 100 h.p. to be reached.

It is perhaps a little unusual that piston type motors may be used for both very low speed, high torque applications and for high speed, high power conditions. Generally radial piston designs satisfy the former requirements and axial piston designs the latter, although the axial piston, Hydro-Titan design of continental origin can give theoretical value of output torque of more than 100 000 Nm.

An alternative type of comparison is provided by *Figure 5.9*, which has been derived from results given by Mortenson.[21] It shows the variation in the prices and power per unit volume of axial piston, variable delivery pumps and fixed delivery vane pumps for automative applications since 1941. Although asymptotic values have not been reached it suggests that whereas future development is unlikely to produce many drastic advances in the design and production of vane type machines there may well be scope for such advances in the axial piston machine field. It should be remembered, however, that these figures come from a very large and highly developed Company in the USA, which probably spends more each year on research and development than the actual annual turnover of many of the smaller Companies in the United Kingdom. The results shown cannot therefore be said to give the general pattern of development, but rather to indicate probable future trends. The factor of five in the price paid for the advantages of a variable delivery unit in terms of both actual cost and power to weight ratio should be noted.

It is beyond the intended scope of this book to present an analysis of, and a comparison between, the pulsations in the flow rates from the various machines but comprehensive treatments of this matter have been given by Faisandier[11] and Grosser.[22]

5.3 Positive displacement pump characteristics

The pumping action of most positive displacement pumps is in general similar to that of vane pumps, and the configuration of the latter may therefore be used in deriving the basic characteristics of such machines. A simplified diagram of a vane pump is shown in *Figure 5.10* and it is assumed to be driven by a constant torque motor.

5.3.1 Flow rate and torque equations

The actual power delivered by the driving motor will be defined as $2\pi N T_a$, where N is its speed and T_a its torque, whilst that delivered by the pump will be $Q_a P$. The actual delivery, Q_a, at the delivery pressure, P, is given by

$$Q_a = Q_g - Q_l - Q_r \qquad (5.1)$$

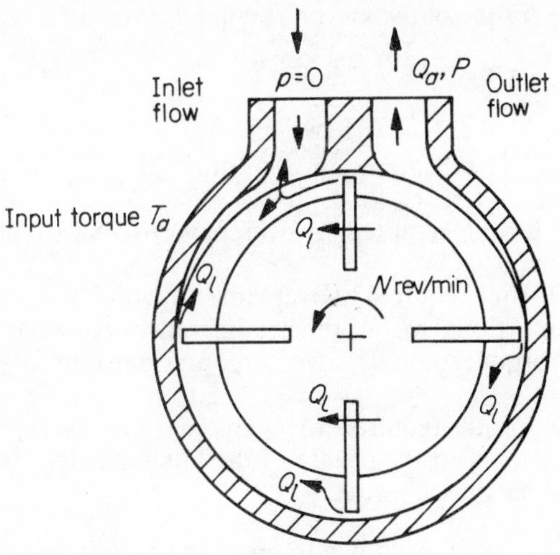

Figure 5.10. Simplified pump configuration

where Q_g is the geometrical or actual swept volume per second of the pump and is given by $2\pi NV$, where V is the swept volume of the machine per radian of rotation, Q_l is the leakage from the delivery to the inlet side, and Q_r is the inlet losses due to inlet restrictions, cavitation or entrained gases.

Bearing these definitions in mind, it follows that

$$Q_g = 2\pi NV \tag{5.2a}$$

and

$$Q_l \propto Ph_1^3/\mu \tag{5.2b}$$

where h_1 is some mean effective clearance path height and μ is the fluid viscosity. It is convenient to refer to the leakage flow rate as the 'slip' flow and to define a 'slip coefficient' C_s by

$$C_s = Q_l/\{2\pi VP/\mu\} \tag{5.3a}$$

remembering that $2\pi V$ is the geometrical swept volume of the machine per revolution of the shaft. Hence combining equation 5.2b and 5.3a we may write

$$C_s \propto h_1^3/V = k_1\,(h_1/D_s)^3 \tag{5.3b}$$

where k_1 is a constant proportional to $D_s^2/2\pi$ and D_s is any convenient representative length such as a piston diameter etc. The actual delivery may be written as

$$Q_a = 2\pi NV - \frac{2\pi VC_s P}{\mu} - Q_r \tag{5.4}$$

The actual torque of the motor T_a will be given by

$$T_a = T_g + T_v + T_s + T_f (+ T_\rho) \tag{5.5}$$

where

- T_g is the ideal torque required to move the piston against the pressure difference P;
- T_v is the torque required to overcome viscous friction in the machine;
- T_s is the torque required to overcome any solid friction (this will be proportional to the pressure difference across the machine);
- T_f is the torque required to overcome any constant solid friction; and
- T_ρ is the torque required to overcome any losses which are proportional to the square of the fluid velocity and may be called the kinetic torque.

The latter term has been shown to occur in practice by Schlösser,[23] but in general it is so small that it can usually be, and from now on, will be neglected.[24] Now

$$T_g = VP \tag{5.6}$$

and since the poiseulle drag on the moving parts is generally small compared with that associated with the couette flow, and is not necessarily in the direction of motion, the viscous drag force F_v is approximately given by

$$F_v = \text{viscosity} \times \text{velocity gradient} \times \text{area} \tag{5.7a}$$

Introducing a drag coefficient C_D given in terms of the viscous torque by

$$C_D = T_v / VN\mu \tag{5.7b}$$

which gives

$$C_D \propto \frac{D_s}{h_1} \tag{5.7c}$$

where D_s is the same representative length used in deriving equation 5.3b.

Similarly a solid friction coefficient may be defined as

$$C_f = \frac{T_s}{PV} \tag{5.8}$$

and substitution of equations 5.6, 5.7b and 5.8 into equation 5.5 then gives

$$T_a = PV + C_D V\mu N + C_f PV + T_f \tag{5.9}$$

5.3.2 Pump efficiency

Defining the volumetric efficiency η_v as Q_a/Q_g gives, from equations 5.1 and 5.4.

$$\eta_v = \frac{2\pi NV - 2\pi VPC_s/\mu - Q_r}{2\pi NV}$$
$$= 1 - C_s P/\mu N - Q_r/2\pi NV \qquad (5.10a)$$

Similarly defining the mechanical efficiency η_m as T_g/T_a gives

$$\eta_m = 1/(1 + C_D \mu N/P + C_f + T_f/PV) \qquad (5.10b)$$

and the overall efficiency η_0 is the product of these two equations, i.e.

$$\eta_0 = \eta_v \eta_m = \left(\frac{1 - C_s P/\mu N - Q_r/2\pi NV}{1 + C_D \mu N/P + C_f + T_f/PV}\right) \qquad (5.10c)$$

For most modern pump and motor designs the inlet loss term Q_r is negligible as is the constant coulomb friction term T_f and as a result the equation may usually be written as

$$\eta_0 = \frac{1 - C_s/S}{1 + C_D S + C_f} \qquad (5.10d)$$

where $S = \mu N/P$ and is a form of Sommerfeld number.

It can be seen that if S varies and $\to C_s$ then $\eta_o \to 0$, and similarly if $S \to \infty$, $\eta_o \to 0$.

Differentiation of equation 5.10d with respect to S shows that the maximum efficiency will occur when

$$S = C_s\left[1 + \sqrt{\left\{1 + \frac{(1 + C_f)}{C_D C_s}\right\}}\right] \qquad (5.11a)$$

and that its magnitude will be given by

$$\eta_{o\,\max} = \frac{1}{1 + C_f + 2C_D C_s\left\{1 + \sqrt{\left(1 + \frac{1 + C_f}{C_D C_s}\right)}\right\}} \qquad (5.11b)$$

Typical theoretical results for tests are shown in *Figure 5.11a* and Schlösser has recently proposed[25] that a further term, dependent on the Euler number should be included in the analysis. The Euler number is equal to the reciprocal of the square root of the product of the Reynolds and Sommerfeld numbers and represents the square root of the ratio of hydrostatic to inertia forces.

5.3.3 Evaluation of loss coefficients

The values of the loss coefficients of a machine may be determined experimentally by performing a series of tests as follows. The input or applied torque, T_a, is measured whilst the pump is run at constant delivery pressure but variable speed. From equation 5.9 it can be seen that the slope of the line will be the coefficient of the speed term, i.e., $C_D V\mu$ and provided care is taken to keep both the pressure and temperature (and hence the viscosity) constant the value of C_D may be obtained from the results of this test as shown in *Figure 5.11*. A further test at constant speed but with variable pressure will give a torque/pressure characteristic (equation 5.9) having a slope of $V(1+C_f)$ and as a result C_f may be found.

If the delivery from the pump during this latter test is plotted against pressure, the slope of the curve will be $2\pi VC_s/\mu$ as shown by equation 5.4 so that C_D, C_f and C_s are then known. Checks on these values may be made from the overall efficiency curve, which intercepts the zero efficiency line at a value of S of C_s, and, which has a maximum value at the value of S given by equation 5.11a.

5.4 Positive displacement motor characteristics

5.4.1 Flow rate and torque equations

The losses of a positive displacement motor are very similar to those of a positive displacement pump and as a result the flow and torque equations 5.1 and 5.5 may be used but with the sign of each term except the first reversed. Hence

$$Q_a = Q_g + Q_l + Q_r \qquad (5.12)$$

and

$$T_a = T_g - T_v - T_s - T_c(-T_\rho) \qquad (5.13)$$

5.4.2 Motor efficiency

Using the same notation and methods as those for deriving the pump efficiency equations gives

$$Q_a = 2\pi NV + \frac{2\pi VC_s P}{\mu} + Q_r$$

and

$$T_a = PV - C_D V\mu N - C_f PV - T_c$$

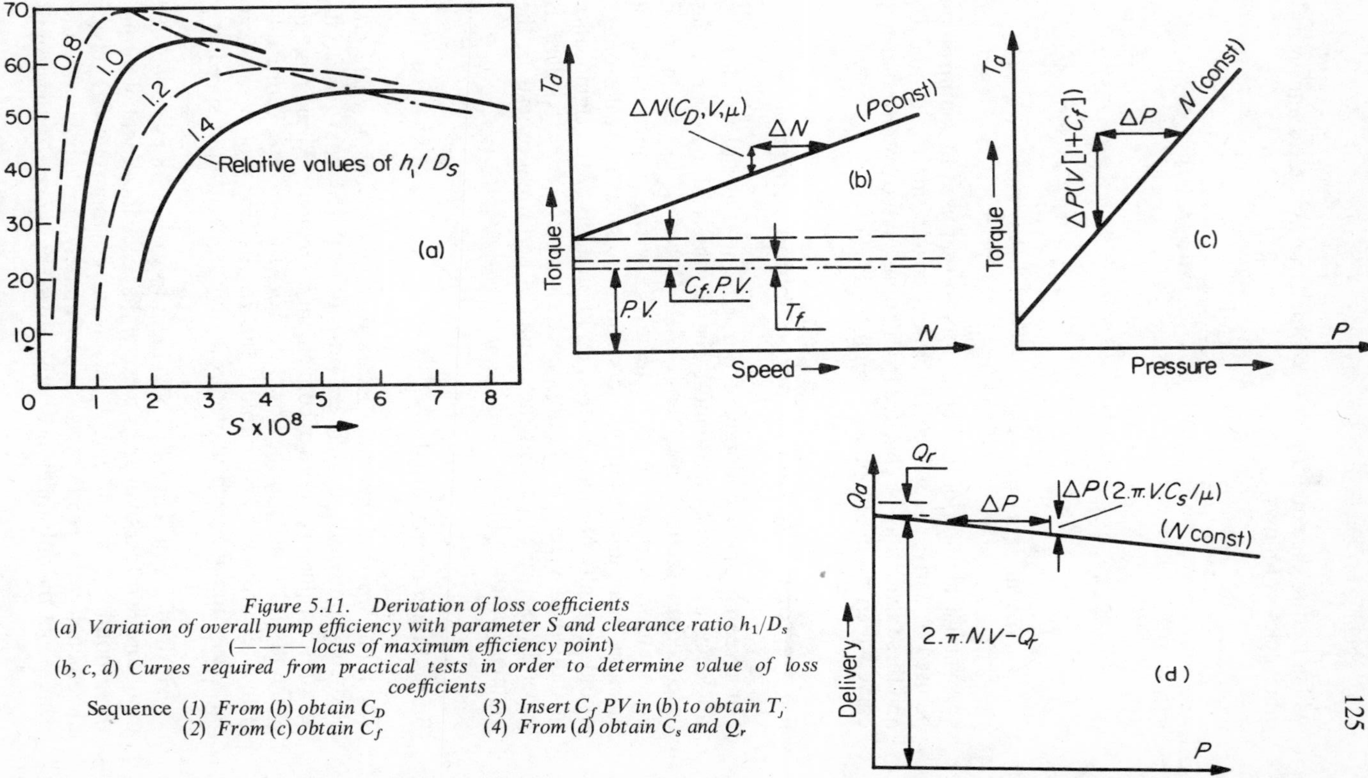

Figure 5.11. *Derivation of loss coefficients*

(a) Variation of overall pump efficiency with parameter S and clearance ratio h_1/D_s (— — — locus of maximum efficiency point)

(b, c, d) Curves required from practical tests in order to determine value of loss coefficients

Sequence (1) *From (b) obtain C_D*
(2) *From (c) obtain C_f*
(3) *Insert C_f PV in (b) to obtain T_j*
(4) *From (d) obtain C_s and Q_r*

which lead to the equations for volumetric, mechanical and overall efficiencies as follows

$$\eta_v = \frac{2\pi NV}{2\pi NV + 2\pi VC_s P/\mu + Q_r}$$

$$= \frac{1}{1 + C_s P/\mu N + Q_r/2\pi NV} \tag{5.14a}$$

$$\eta_m = 1 - C_D \mu N/P - C_f - T_c/PV \tag{5.14b}$$

and

$$\eta_o = \frac{1 - C_D \mu N/P - C_f - T_c/PV}{1 + C_s P/\mu N + Q_r/2\pi NV} \tag{5.14c}$$

A similar method of testing motors to that used for pumps may be used to derive the values of the loss coefficients C_D, C_s and C_f and these again may be checked from points on the overall efficiency curve which should be plotted against values of S.

5.5 Performance testing and machine controls

For research and development purposes it is useful to plot the efficiency of a machine against the Sommerfeld number, S, as described in sections 5.3.3 and 5.4.2. However, for commercial presentation, it is obviously better to show the performance of a machine in a more readily understood form since most customers would generally not have a detailed knowledge of the significance of individual values of the loss coefficients, etc. As a result it is general practice to present the results of performance tests in the manner described below.

5.5.1 Pump characteristics

The input speed of the pump is set at some convenient value and held constant whilst the delivery pressure is raised from zero to its maximum value in a series of convenient increments. Measurements of the input torque and the flow rate from the unit are recorded at each pressure level and initially the results are plotted as flow rate—*v*—pressure, each line representing a given input speed, say N_1, N_2, N_3, etc.

For simplicity it is usually convenient to work with increments of input torque rather than pressure such that each increment of torque corresponds to a convenient increment of input power. These increments will correspond approximately to equal pressure increments, the only difference being caused by changes in the overall efficiency with pressure. If these incremental values of torque are T_1 and $N_1 = N_2/2 = N_3/3$, etc., then the input power at the speed N_1 and a torque of $2T_1$ is equal to that at a speed N_2 and a torque T_1, etc., so that lines of constant input power may be superimposed on the flow rate/pressure characteristics. They will be approximately hyperbolic in form as shown in *Figure 5.12a*.

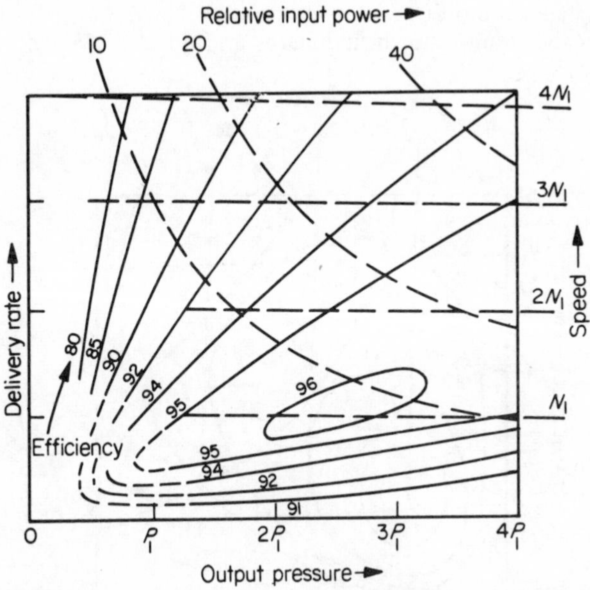

Figure 5.12(a). General pump performance characteristics

Contours of constant overall efficiency are then added as shown and the closed ring form of their shape follows from equation 5.10c and from the more general curve in *Figure 5.11*.

The form of these results is most convenient to the user of such a machine since they provide values of practical quantities with which he will be familiar and which he will wish to know in order to assess whether or not the machine will be suitable for the application he has in mind.

5.5.2 Motor characteristics

For motor tests it is usual to keep the supply pressure constant and plot output torque against output speed. It is again convenient if increments of speed are chosen such that equal increments of output horsepower are represented by them, but care must be taken to ensure that allowance for the fall off in the output torque with speed is made. Constant output horsepower curves may then be superimposed as before and again these are approximately hyperbolic in form.

Contours of constant efficiency may be drawn in and an actual set of results for a motor are shown in *Figure 5.12b*.

Both the results shown in *Figures 5.12a* and *5.12b* are for axial

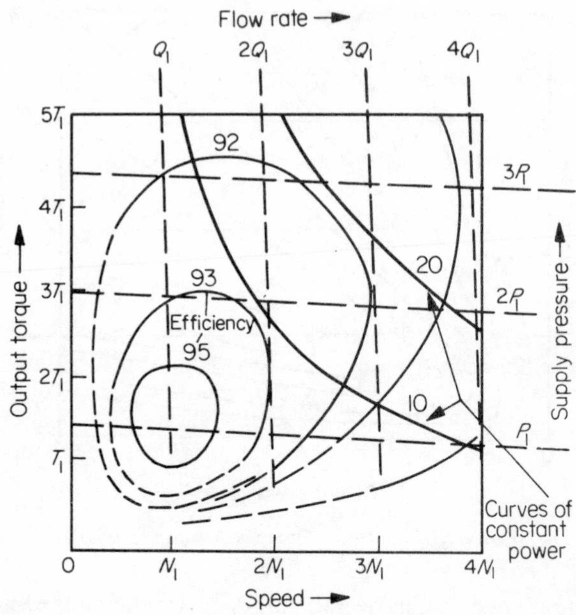

Figure 5.12(b). General motor performance characteristics

piston type machines of identical size and they are naturally very similar in appearance. Details of suitable test facilities are given by Platt and Kelly[26] who have paid particular attention to the effects of the use of fire resistant fluids on pump and motor life.

5.5.3 Pump control

If a fixed delivery pump is run at constant speed and the flow rate demand varies, then the simplest but most wasteful means of providing a flow control is to utilise the relief valve as a 'blow-off' valve so that as the flow demand decreases some of the delivery from the pump is returned to the tank via the relief valve. Such systems have little to recommend them except for the low initial cost since the cost of the power wasted, particularly at zero and low flow demands, will rapidly mount and the system is very uneconomical.

It is sometimes argued that for low power units the cost of providing a variable delivery pump and flow control is often unjustified but even this is seldom the case as the following example will show. Consider a 5 kW unit which will be operating for a nominal 40 hour week usefully working at an average power equal to its maximum for, say, 20 hours per week. If the power cost is, say, 1p/kW hour then in a week the cost of the wasted power is approximately £1 and in a year this will amount to £50. Allowing for the more rapid wear of the unit when working continuously on full load and therefore its more frequent replacement and, in addition, for the more rapid degredation of the hydraulic fluid due to its continual shearing on passing through the relief valve and its subsequently more frequent replacement, it is easy to follow why controlled delivery systems rather than 'blow-off' systems are generally more economical.

The most commonly required type of pump control is one which maintains the delivery pressure at some predetermined value irrespective of the flow demand. The simplest device, incorporating a variable delivery pump, is one in which the delivery pressure is fed onto one side of a ram or piston, the motion of which is opposed by a spring. The spring force may be varied to correspond to the required delivery pressure and the ram is mechanically linked to the pump swashplate or eccentric. If the flow demand drops and the pressure therefore rises, this increase in pressure on the ram compresses the spring and thereby adjusts the ram position and hence the pump swashplate angle to correspond to the lower flow demand.

A more complicated system is one in which the ram is replaced by a spool valve and this in turn governs the position of a jack which is coupled to the swashplate. A feedback link is required between the jack and valve and this for example can be achieved by having a floating sleeve valve such that movement of the jack gives rise to movement of the valve sleeve thereby shutting off the valve. An actual example of this type of pressure control unit is shown in *Figure 5.13*, and is described in detail by Price.[27] It is arranged such that it may operate for both positive and negative values of swashplate angle and an analysis of the response of such a system is given by Rigby.[28] For large units it is not only desirable

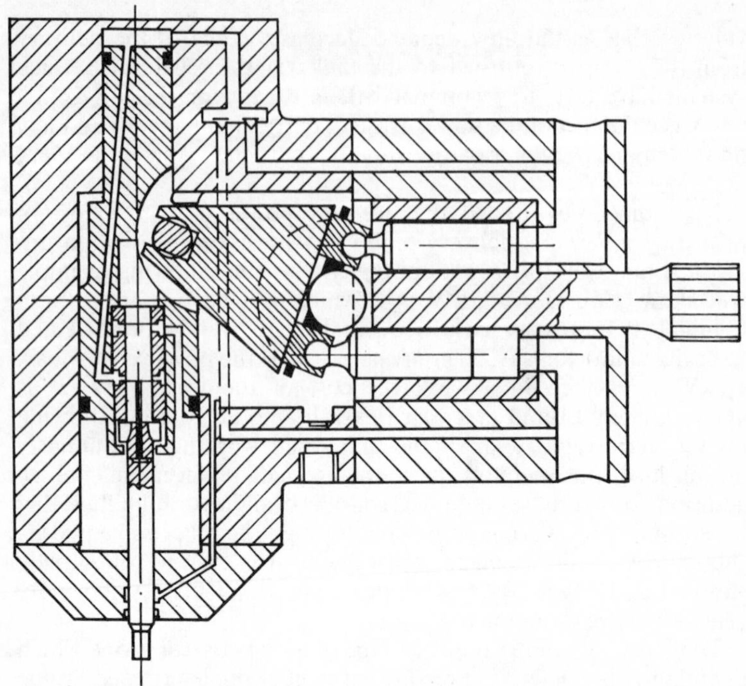

Figure 5.13. Pressure control system

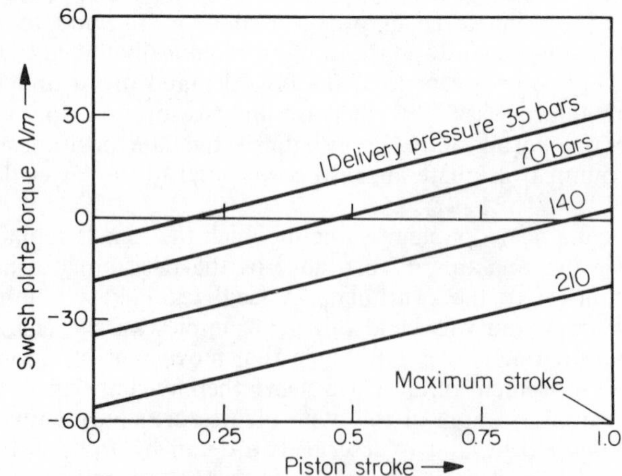

Figure 5.14. Variation of torque required to tilt swashplate with piston stroke (delivery) and delivery pressure (constant speed)

to incorporate this type of control to maintain constant pressure but also to avoid undue effort if manual operation is envisaged since the couple required to change the delivery during operation can be considerable as is illustrated in *Figure 5.14*.

It should be noted that constant pressure pump systems are a form of automatic control system or servomechanism in which the input is the desired pressure (this is generally 'fed' into the system by the

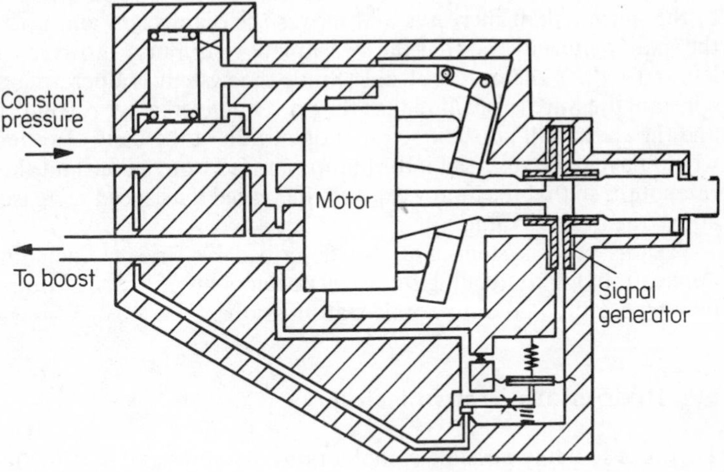

Figure 5.15. Constant-speed aircraft drive supplied with constant pressure

pressure setting lever) and the output is the actual delivery pressure. The feedback link is provided by the pressure regulating device which compares the values of the desired and actual delivery pressures and then makes a correction accordingly. The 'load' on the system is represented by the changes in demanded flow rate and the response may be judged by the speed with which the output pressure responds to step and sinusoidal variations in demanded flow.

5.5.4 Motor control

Whilst most pump controls are designed to ensure that the delivery pressure of the machine remains constant under fluctuating flow rate demands most hydraulic motor controls are designed to ensure that the motor speed remains constant under fluctuating load conditions at the motor output shaft. As a result a speed sensing device is generally required which emits a signal proportional to the output shaft speed which is then compared with the desired output speed and any difference between these two quantities is applied to the motor swashplate to correct the error.

The motor is fed with either a constant flow rate or a constant pressure and a device, used in aircraft applications, is shown in *Figure 5.15*. If the speed of motor rises above its desired value then the pressure generated by the small centrifugal pump unit attached to the output shaft increases and moves the diaphragm and opens the small plate valve so that the swashplate piston may move to the left and reduce the motor stroke. Since the system is operating at constant pressure this will reduce the power delivered per revolution and the speed will fall. The sense of operation of the control system would have to be reversed if the pump were fed from a constant flow rate source so that the motor capacity increased if its speed increased above the desired value.

Various other systems are also described by Price,[27] including applications for controlling the acceleration of lifts, fan drive control by temperature and automobile transmissions.

5.6 Hydrostatic transmissions

There are so many possible combinations of pumps and motors that may be chosen to give satisfactory variable speed drives or transmissions that it is beyond the scope of this book to consider even those which have been successfully developed and used. However, further details appear in the Proceedings of the 'Conference on Oil Hydraulic Power Transmission and Control', third session (I.Mech.E., November 1961) where papers were presented by Bowers[29], Westbury *et al*,[30] Price[27] and Constantinesco.[31]

The opening remarks at this session, made by Professor H. Thoma, give interesting details of the application of hydrostatic transmissions to shunting locomotive drives, tractors and the main propulsion unit and dredging drive of a sea-going dredger. Bower's paper is also of particular interest and was mainly concerned with automotive applications. Other work on this topic appears in References 32 and 33, the latter also including a useful bibliography. More recent work is described in Reference 34.

The output power and torque curves of three commonly used systems are shown diagrammatically in *Figure 5.16*. The input speed to the pump is assummed to be constant in each case and with a fixed delivery pump and variable stroke motor the operating characteristics will be as shown in *Figure 5.16a*. The other two

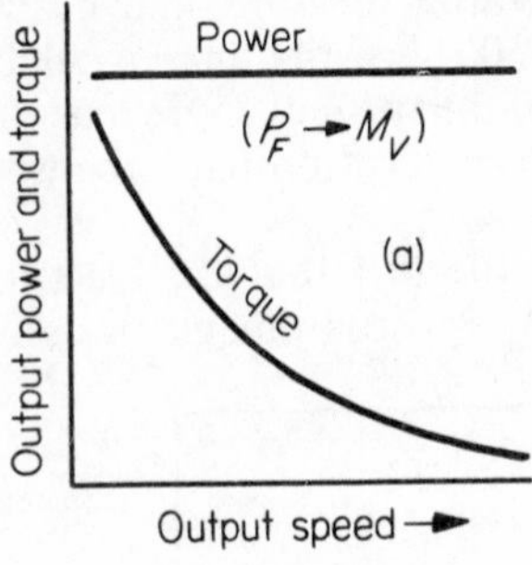

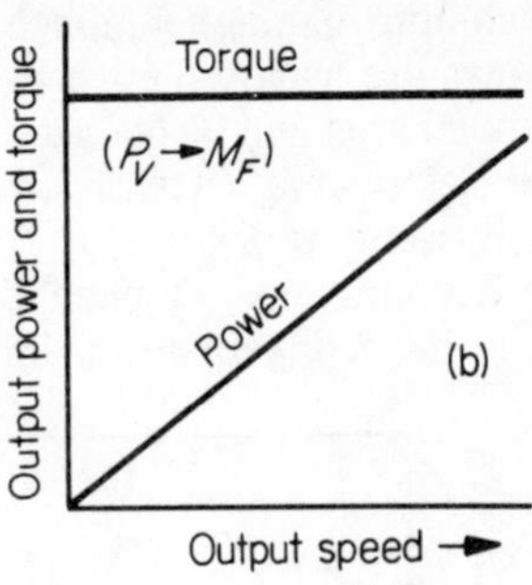

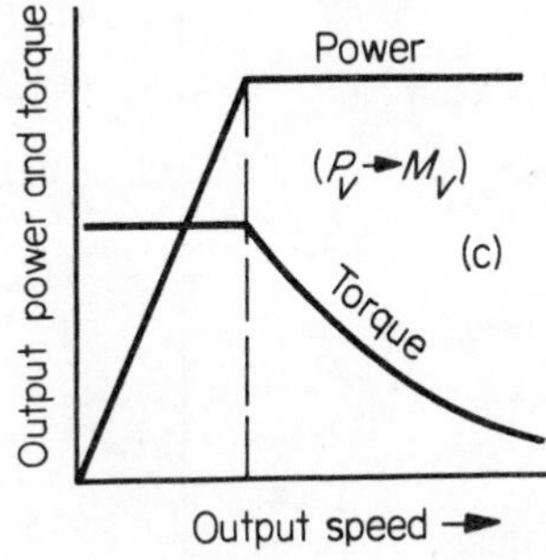

Figure 5.16. Variation of output power and torque with output speed for three types of hydrostatic transmissions

P. Pump *M. Motor*
F. Fixed stroke *V. Variable stroke*

diagrams are for a variable delivery pump and fixed stroke motor combination and for a variable delivery pump and variable stroke motor combination respectively and illustrate the wide range of performance obtainable with hydrostatic machines which in turn partly accounts for their use in an increasing number of fields. Other attractions are their advantages over electric motors in terms of high torque to inertia ratios, giving high speeds of response and low time constants, wide speed ranges, low weight/power ratio, satisfactory performance at very low speeds (absence of stick-slip effects) and relatively small stopping and reversal times.

Several of these advantages can in some way be attributed to the inherent disadvantage suffered by electric motors in terms of their 'equivalent operating pressure'. If a hydraulic ram and a tractive electromagnet are considered then using good electrical steels in the latter it will be found that the saturation density will be about 250 maxwells/mm^2 which is equivalent to a force of about 1.5 N for each mm^2 of magnet area. Since operating pressures in hydraulic machines are usually between 70 and 500 bars, the force available for a given area will be between four and thirty times as great as that of the electro-magnet with an average value of probably between ten and twenty times.

No account has yet been taken of the fact that the electrical machine has stringent design limitations imposed by magnetic design

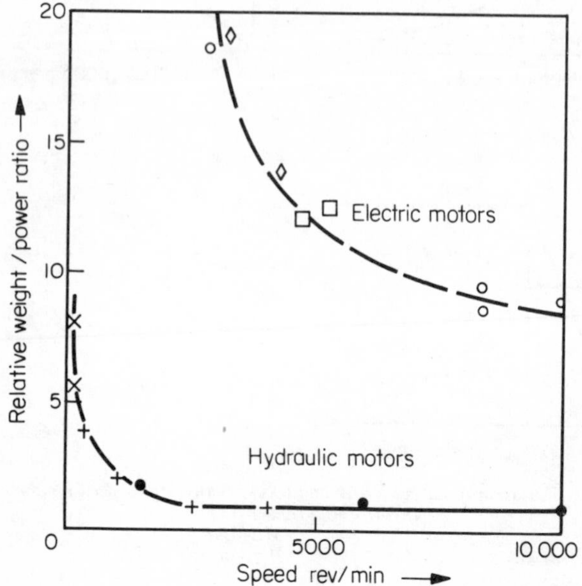

Figure 5.17(a). Variation of weight to power ratio with maximum speed for electrical and hydraulic motors

requirements and as these do not exist for the hydraulic design a far wider choice of materials is available. Blackburn *et al*[35] suggest that for large machines the torque to inertia ratio of a hydraulic machine could be several thousand times greater than that of an equivalent electrical one, and the actual variation of the weight/power ratio with speed for both hydraulic and electrical machines is shown in *Figure 5.17a*.

5.6.1 Response of hydrostatic transmission

Hydrostatic transmissions have been used for many years as the basic power element of several designs of hydraulic servomechanisms and position control systems and in particular in 'fire-control' systems for both naval and anti-aircraft applications. As a result considerable work has been done on deriving the response of the motor output shaft position to changes in the pump swashplate angle, which represents the input to the control system. A simplified diagram of such a transmission is shown in *Figure 5.17b* and if Q_g is the ideal or geometrical delivery from the pump and Q_m that to the motor then

$$Q_g = Q_m + Q_l + Q_c \tag{5.15}$$

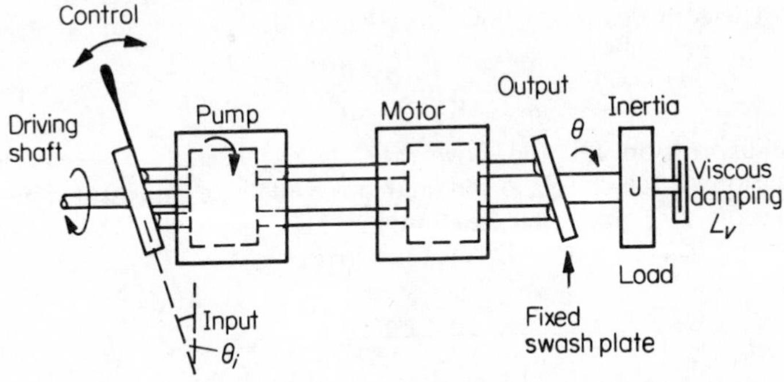

Figure 5.17(b). Hydrostatic transmission system having inertial and viscous loads

where Q_l and Q_c represent the total leakage losses of the pump and motor and the compressibility flow respectively and the flow loss due to restrictions is ignored. The ideal displacement of the pump is given by

$$Q_g = K_p \theta_i \tag{5.16a}$$

where θ_i is its swashplate angle and K_p is its delivery per unit of swashplate tilt and the actual velocity $\mathrm{d}\theta/\mathrm{d}t$ of the motor output shaft will be given by

$$Q_m = K_m \,\mathrm{d}\theta/\mathrm{d}t \tag{5.16b}$$

where K_m is the volumetric displacement of the motor.

The leakage flow will be given by

$$Q_l = PK_L \tag{5.16c}$$

where P is the pressure drop across the system and K_L is a leakage coefficient. The motor torque PK_m will be equal to the sum of the torques required to accelerate the inertial load J and to overcome the viscous load $L_v \,\mathrm{d}\theta/\mathrm{d}t$.

Substituting for P in equation 5.16c gives

$$Q_l = \frac{K_L J}{K_m}\left(\frac{d^2\theta}{dt^2}\right) + \frac{K_L L_v}{K_m}\left(\frac{d\theta}{dt}\right) \tag{5.16d}$$

If V is the volume of oil under compression then the change in its volume δV when it is subjected to the pressure P is given by

$$\delta V = \frac{V}{K^1} P$$

where K^1 is the equivalent bulk modulus of the fluid allowing for any free air in the system and the elasticity of the connecting pipes, etc. The compressibility flow rate $\delta V/\delta t$ is therefore given by

$$Q_c = \frac{\delta V}{\delta t} = \frac{V}{K^1 K_m}\left(\frac{J\, d^3\theta}{dt^3} + \frac{L_v\, d^2\theta}{dt^2}\right) \tag{5.16e}$$

The expression for the flow rates $Q_g\, Q_m\, Q_l$ and Q_c given by equations 5.16a, 5.16b, 5.16d and 5.16e may be substituted into equation 5.15 and after rearranging and using the D-operator notation where $d\theta/dt = D\theta$ etc., this gives

$$\frac{\theta}{\theta_i} = \frac{K_p/K_m}{D\left\{\dfrac{VJ}{K^1 K_m^2} D^2 + \left(\dfrac{VL_v}{K^1 K_m^2} + \dfrac{K_L J}{K_m^2}\right) D + \left(\dfrac{K_L L_v}{K_m^2} + 1\right)\right\}} \tag{5.17a}$$

The presence of the quadratic term in D in the denominator shows that under certain conditions the system may have an oscillatory response but Chestnut and Mayer[36] suggest that the frequency is usually fairly high (5 to 10 Hz). The factor of D outside the bracket shows that systems act as an integrator at low frequencies and if the value of V/K^1 is very small such that the two terms in which it appears can be neglected in comparison with the others then

$$\frac{\theta}{\theta_i} = \frac{K_p/(K_m + K_L L_v/K_m)}{D(T_{Hl} D + 1)} \tag{5.17b}$$

where the time constant T_{Hl} is given by

$$T_{Hl} = \frac{J K_L}{(K_m^2 + L_v K_L)}$$

If no viscous load is present, $L_v = 0$, so that

$$\frac{\theta}{\theta_i} = \frac{K_p/K_m}{P(T_{H2} D + 1)} \tag{5.17c}$$

Where $T_{H2} = JK_L/K_m^2$ and this is similar in form to the response equation for the shaft position of a d.c. electric motor subjected to a change in the applied voltage.[36] Further aspects of the effects of system elasticity and compressibility have been studied by McCallion *et al.*[37]

5.7 Some design considerations

The performance of most positive displacement machines is very dependent on the satisfactory operation of one or more key components. In the vane pump for example it is essential that the vanes are satisfactorily balanced such that they neither bear too heavily against the casing, giving rise to excessive wear and damage, nor should they be forced away from the casing by a hydrodynamic lifting force generated in the fluid film between the vane tip and the casing, thereby causing an excessive leakage loss. The critical thrust balancing of the end plates in gear pumps is described in detail in Reference 3, but there is surprisingly little published information on design techniques for either gear or vane machines.

In the case of piston machines, there are usually several regions where particular care must be taken if a satisfactory design is required and, in recent years, a certain amount of analysis has at least provided a basis for the preparation of design data.[18,19]

5.7.1 Losses of axial piston machines

A simplified diagram of an axial piston machine is shown in *Figure 5.18a* where it can be seen that rotation of the shaft will cause the rotor or cylinder block to rotate which, in turn, will cause the pistons to reciprocate in their cylinders. Under pumping conditions, contact between the slipper pad bearings attached to the pistons and the stationary swashplate can be maintained during the suction stroke by the use of a small positive suction pressure, or springs placed between the piston crowns and the cylinder heads or by a spider plate bearing onto the upper surface of the slipper pads. The valve plate is stationary and as each piston moves up its cylinder it forces high pressure oil into the kidney shaped delivery port in the stationary valve plate. During their down strokes they receive low pressure oil from the suction port of the valve plate.

The losses in such machines can be classified in the manner described in section 5.2.3 but it is possible to make a more detailed study of some of these. A common loss condition arises when there is relative motion between two surfaces separated by a small clearance or gap and subjected to a pressure gradient along the gap

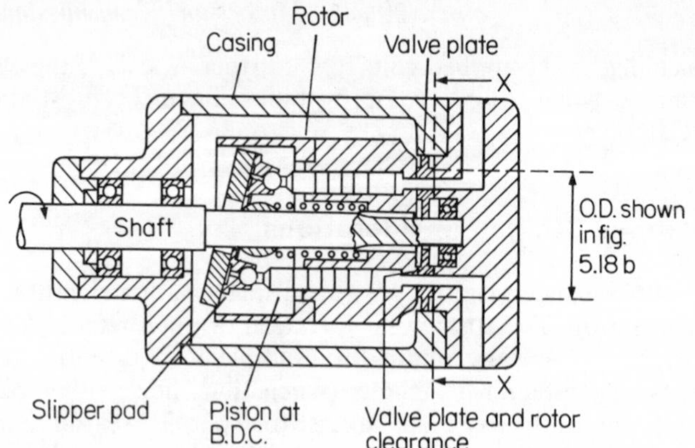

Figure 5.18(a). Simplified configuration of an axial piston machine (Note that the rotor has been rotated by 90° to show pistons at T.D.C. and B.D.C.)

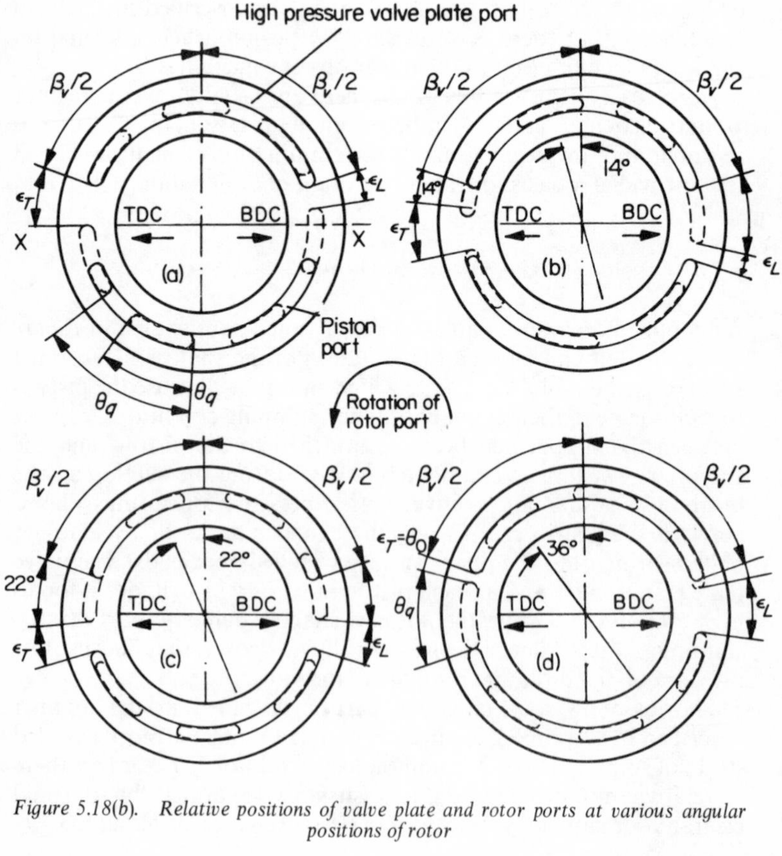

Figure 5.18(b). Relative positions of valve plate and rotor ports at various angular positions of rotor

as shown in *Figure 5.19*. The total power loss of such a system will be given by the sum of the leakage flow rate along the gap multiplied by the pressure drop and the viscous drag force acting on the adjacent faces multiplied by the relative velocity. This may be written as

$$E = PQ_1 + F_v u_r$$

where E is the power loss;
P is the pressure drop;
Q_1 is the leakage flow rate;
F_v is the viscous drag force; and
u_r is the relative velocity of the two surfaces.

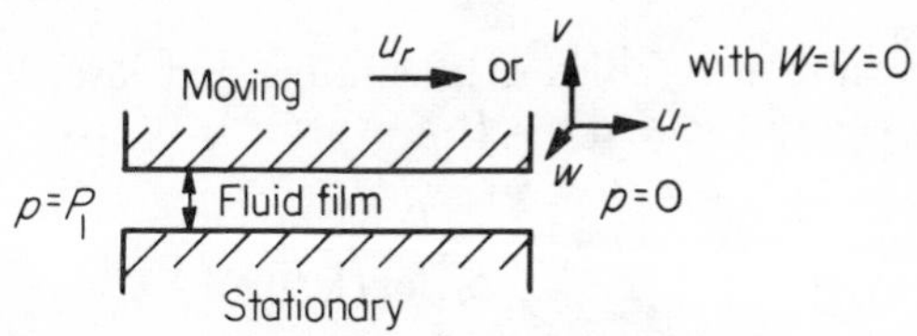

Figure 5.19. Common loss configuration

This method of approach may be applied to the losses occurring between each slipper pad bearing face and the swashplate surface, the losses between the adjacent faces of the rotor and stationary valve plate and to the losses past the pistons and their cylinder walls[18,19] as shown in the following section.

5.7.2 Analysis of slipper pad losses

The method of obtaining an analytical expression for the losses and the condition for which their sum is a minimum is probably best described by considering the simplified analysis of the losses of the slipper pad bearing shown in *Figure 5.20*.

The leakage flow rate past the sill of the pad may first be determined since if P_r is the pressure in the pad recess then considering the flow rate across a circumference of radius r on the face of the pad such that $D_1/2 \leqslant r \leqslant D_2/2$ gives

$$Q_1 = -\frac{\delta p h^3 2\pi r}{\delta r \cdot 12\mu} \tag{5.18a}$$

where the symbols are as shown in *Figure 5.20*, μ is the viscosity and

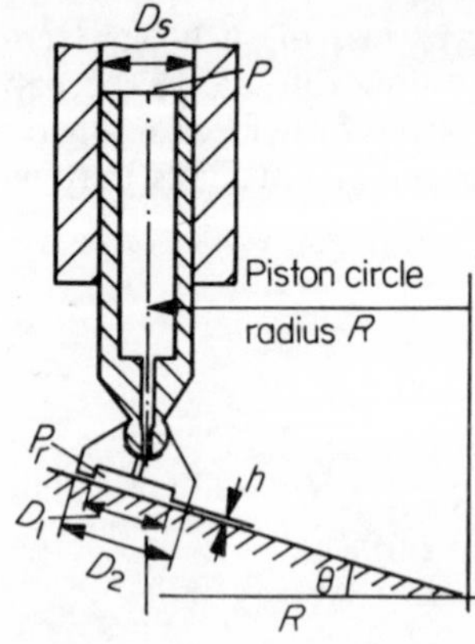

Figure 5.20. Simplified geometry of slipper pad

p is the pressure at a radius r. Integrating and inserting the limits $r = D_1/2$ when $p = P_r$ and $r = D_2/2$ when $p = 0$, gives

$$Q_1 = \frac{P_r \pi h^3}{6\mu \log(1/\bar{D})} \quad (5.18b)$$

where $\bar{D} = D_1/D_2$ and if r_1 is the radius of the choke feeding the recess with the operating fluid this flow rate is also given by

$$Q_1 = \frac{(P - P_r)\pi r_1^4}{8\mu L_1} \quad (5.18c)$$

where L_1 is the choke length and P is the operating pressure. Retruning to the integration of equation 5.18a, the pressure p at any radius r can be written in form

$$p = P_r \left\{ \frac{\log(D_2/2r)}{\log(1/\bar{D})} \right\}$$

if equation 5.18b is used to eliminate the Q_1 term. Multiplying this pressure by the area of an elemental ring given by $2\pi r\,\delta r$, integrating between the limits $D_2/2$ and $D_1/2$ and adding $\pi P_r D_1^2/4$ to account for the force of the pressure in the recess finally gives

$$\text{Force perpendicular to pad face} = \frac{\pi P_r (D_2^2 - D_1^2)}{8 \log(D_2/D_1)}$$

If this expression is multiplied by the cosine of the swashplate angle, θ, the balance of the vertical forces acting on the piston in the absence of friction may be written in the form

$$\text{Vertical force } F_p = \frac{\pi P_r (D_2^2 - D_1^2) \cos\theta}{8 \log(1/\bar{D})} = \frac{\pi P D_s^2}{4} \quad (5.19a)$$

where D_s is the piston diameter,

$$\text{or } P_r = \frac{2P\log(1/\bar{D})\sec\theta}{\bar{D}_2^2(1-\bar{D}^2)} \tag{5.19b}$$

where $\bar{D}_2 = D_2/D_s$.

Since the pistons are only pressurised for approximately half of each revolution, the leakage power loss is given by $\frac{1}{2}PQ_1$ and substituting the value for P_r from equation 5.19b into equation 5.18b and multiplying both sides by $P/2$ then gives

$$\text{Leakage power loss} = \frac{P^2\pi h^3\sec\theta}{6\mu\bar{D}_2^2(1-\bar{D}^2)} \tag{5.20a}$$

For simplification it will be assumed that the slipper describes a circle of radius R on the swashplate face (a more accurate expression for the path of the pad is given in Reference 18) and if the speed of rotation of the machine is N rev/min the shaft power required to overcome the viscous drag between the adjacent faces of the slipper pad and swashplate is given by

$$\text{Drag power} = \frac{\pi^3N^2R^2D_2^2(1-\bar{D}^2)\mu}{3600h} \tag{5.20b}$$

The two components of the total loss have been derived for one piston-slipper combination and this loss must therefore be compared with the power of one piston, i.e. $P\pi D_2^2R\tan\theta N/120$. The efficiency loss E_1 is given by the sum of equations 5.20a and 5.20b divided by the piston power and this may be written as a percentage loss in the form

$$E_1 = 100\bar{h}\left\{\frac{20}{\sin\theta}\beta_s\bar{h}^2 + \frac{\pi^2}{30\tan\theta\beta_s\bar{h}^2}\right\} \tag{5.21}$$

where $\bar{h} = h/D_s$, $\beta_s = (P/\mu N)/\bar{R}\bar{D}_2^2(1-\bar{D}^2)$ and $\bar{R} = R/D_s$.

The expression $\beta_s\bar{h}^2$ is a form of Sommerfeld number divided by the radius ratio and slipper sill area ratio terms. The slipper sill area ratio term $\bar{D}_2^2(1-\bar{D}^2)$ is of particular interest, representing pad bearing area/piston area and is referred to again in section 5.7.5. For most machine designs the terms contained in the slipper parameter β_s will be fixed by operational and minimum bearing area requirements. Under such conditions equation 5.21 may be differentiated with respect to $\bar{h}$ and equated zero, thereby giving the value of $\bar{h}$ which results in the minimum value of the loss. Its value is given by

$$\bar{h} = 0.274/\beta_s^{1/2}(\cos\theta)^{1/4}$$

and since θ is generally limited between 0 and 20° a sufficiently accurate result is given by

$$\bar{h} = 0.274/\sqrt{\beta_s} \tag{5.22a}$$

Alternatively under other design conditions it is possible for a limit to exist on the minimum acceptable value of h and hence one starts with a given value of $\bar{h}$. The alternative approach of differentiating equation 5.21 with respect to β may then be used, giving

$$\beta_s = 0.129/\bar{h}^2 \tag{5.22b}$$

for minimum loss conditions when $\bar{h}$ is fixed.

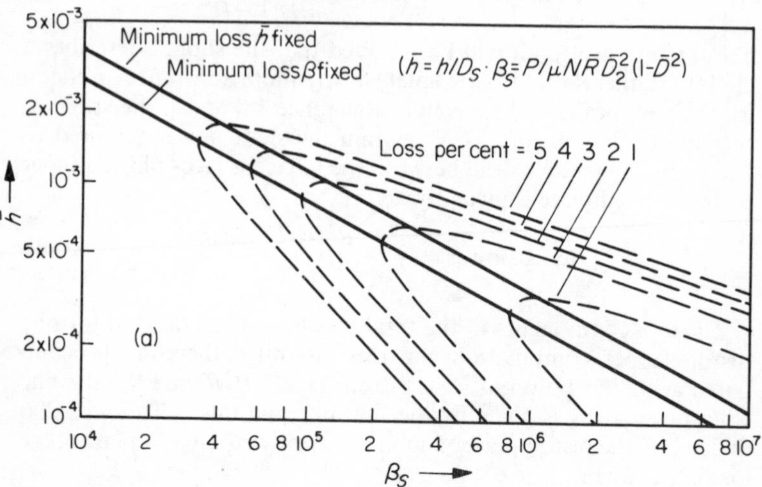

Figure 5.21(a). Variation of film thickness ratio with slipper parameter for minimum loss conditions (swash angle = 10°)

To satisfy equation 5.22a it is possible to adjust h and hence $\bar{h}$ by changes in the radius and/or length of the choke feeding the slipper recess with oil and to satisfy equation 5.22b the value of β_s may be varied by changing the inner and/or the outer diameter of the pad. For the former condition the viscous drag loss is three times that of the flow loss whilst for the latter condition, when β_s is the variable and $\bar{h}$ is fixed, the components of the total loss are equal. A method of presenting these results is shown in *Figure 5.21a* for a swashplate angle of 10° and curves of constant efficiency loss have been added. Since the approximation $\sin\theta \approx \tan\theta$ is valid for most practical values of the swashplate angle the form of the curves of constant efficiency loss are virtually unchanged when θ varies but

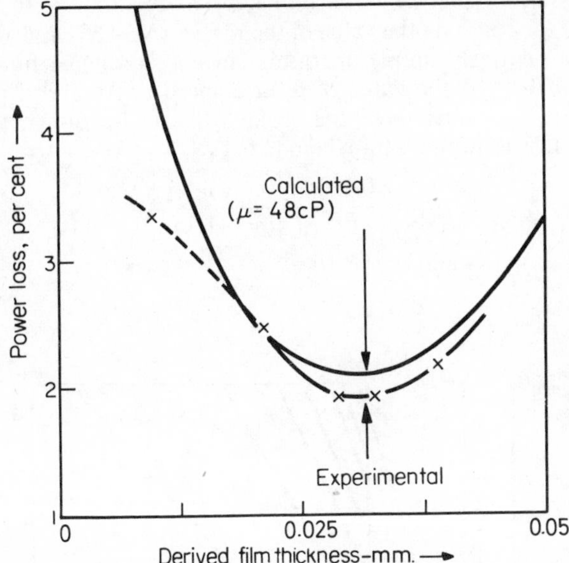

Figure 5.21(b). Experimental variation of power loss at 1000 rev/min with film thickness

the actual values of the percentage loss should be multiplied by (sin 10°)/(sin θ_1) where θ_1 is the appropriate value of the swashplate angle. The experimental variation of the power loss of a piston-slipper pad combination with film thickness is shown in *Figure 5.21b* and illustrates the importance of working with the predetermined value of the latter. The difference between the two curves shown for low values of h is attributed to changes in the effective value of the viscosity at the higher rates of shear and is considered in section 5.7.7.

5.7.3 Effects of speed and pressure ranges on minimum loss conditions

If a machine is to operate over a speed range of N_1 to N_2 rev/min, a pressure range P_1 to P_2 and there is no indication of the conditions under which it will be running for the maximum period of its use then the slippers may be designed such that the average loss for all possible combinations of the values of the speed from N_1 to N_2 and of the pressure from P_1 to P_2 is a minimum. This may be done by integrating equation 5.21 with respect to both the speed, N, and the pressure, P, between the limits N_1 to N_2 and P_1 to P_2 and dividing the result by the value of the ranges $(N_2 - N_1)$ and $(P_2 - P_1)$. This is a relatively simple operation since β_s is a linear function of P/N and if β_{s1} is the value of β_s at a speed of N_1 and a pressure of P_1 and $\bar{h}$ is considered the variable it can be shown that the average loss is a minimum when

$$\beta_{s1} = \frac{0.129}{\bar{h}_2} \sqrt{\left\{\frac{(n^2-1)\log p}{(p^2-1)\log n}\right\}} \tag{5.22c}$$

where $n = N_2/N_1$ and $p = P_2/P_1$.

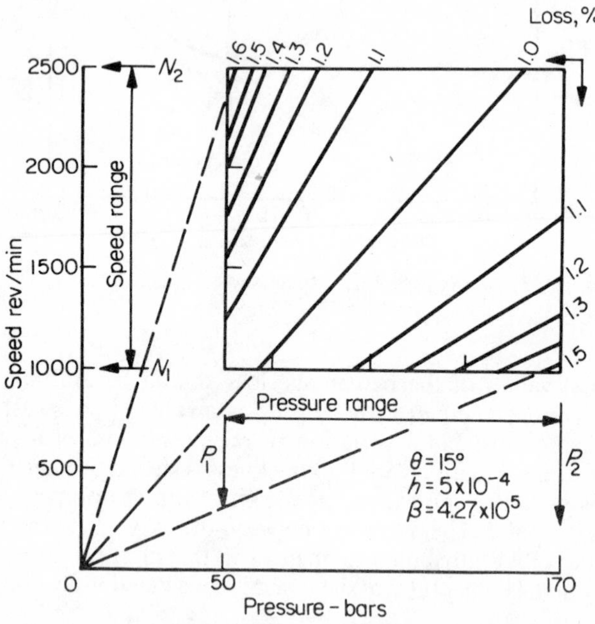

Figure 5.22. Variation of power loss over speed and pressure ranges (minimum average loss conditions)

As an example of the application of equation 5.22c, consider the design of a pad suitable for a machine having a speed range of 1000 to 2500 rev/min, a pressure range of 50 to 170 bars and a given minimum gap ratio $\bar{h}$ of 5.10^{-4}. Then $n = 2.5$, $p = 3.4$ and the value of $\beta_{s1} = 4.27 \cdot 10^5$ from equation 5.22b. The power loss at any value of p and n is obtained by substituting the corresponding value of β_s (i.e. $\beta_{s1} PN_1/P_1 N$) into equation 5.21. It follows that since the efficiency loss is a function of the ratio P/N its value will be constant if this ratio is constant and as a result there is a straight line relationship between P and N for any given value of power loss.

Figure 5.22 shows the variation of the 'minimum average' efficiency loss for a speed range of 1000 to 2500 rev/min and a pressure range of 50 to 170 bars.

5.7.4 Piston and valve plate losses

As described in section 5.7.1 the drag and flow losses of both the pistons and valve plate of axial piston machines are of the same form as those of the slipper pads and may therefore be analysed

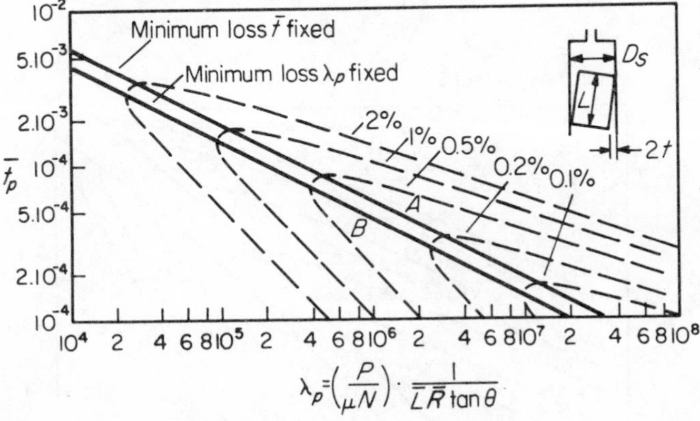

Figure 5.23. Variation of clearance ratio with piston parameter for minimum loss conditions (piston fully tilted)

by similar methods. Results of such analyses are given in Reference 19 and have been summarised in *Figures 5.23* and *5.24*. The former shows the conditions of minimum loss for pistons when they are fully tilted in their cylinders in terms of a piston parameter λ_p and the radial clearance ratio $\bar{t}_p$.

The efficiency loss equation may be written in the form

$$E_p = 100\bar{t}_p\left\{8.75\lambda_p\bar{t}_p^2 + \frac{\pi^3}{30\lambda_p \cdot \bar{t}_p^2}\right\} \tag{5.23}$$

where E_p is the percentage power loss, $\bar{t}_p = t/D_s$, $\lambda_p = (P/\mu N)/\bar{L}\bar{R}\tan\theta$, $\bar{L} = L/D_s$ and $\bar{R} = R/D_s$ where R is the piston pitch circle radius and the notation is as shown in *Figure 5.23*. The piston operating

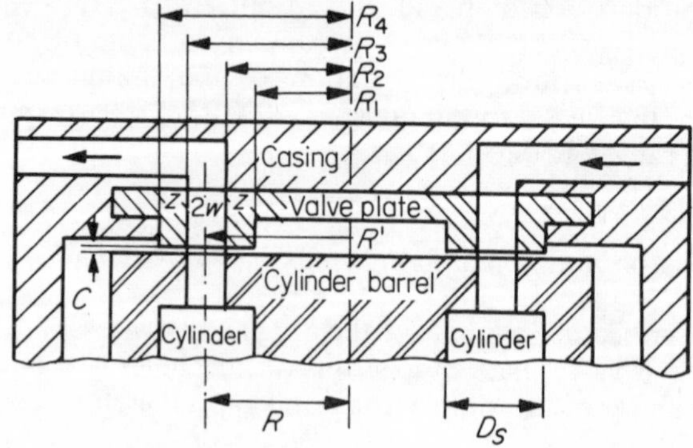

Figure 5.24(a). Simplified valve plate configuration

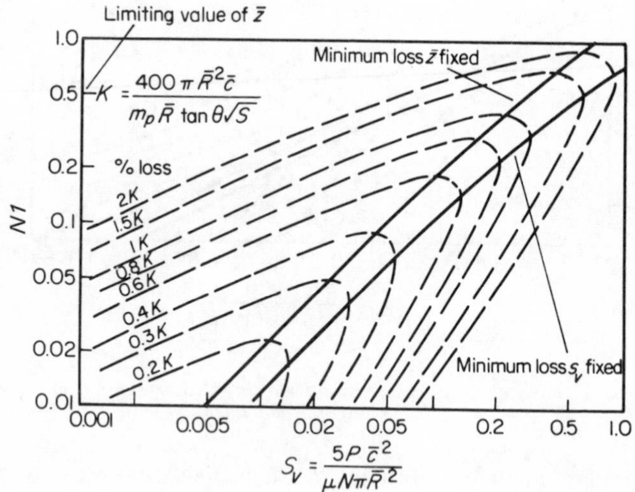

Figure 5.24(b). Variation of land width ratio with valve plate parameter for minimum loss conditions

parameter is seen to be similar to that of the slipper pad β insofar as it has the form of a Sommerfeld number when multiplied by $\bar{t}_p^2$, the square of a clearance ratio.

The simplified valve plate configuration shown in *Figure 5.24a* has been used to derive the minimum loss conditions shown in *Figure 5.24b* in terms of the width ratio of the valve plate lands, $\bar{z}$, given by $\bar{z} = z/R$ and the valve plate operating parameter S_v given by

$$S_v = 5(P/\mu N)\bar{c}^2/\pi(\bar{R})^2 \tag{5.24}$$

where $\bar{c} = c/D_s$, $\bar{R} = R/D_s$ and the dimensions are as shown in *Figure 5.24a*. It will be noted that a Sommerfeld group also appears in the value of the parameter. The actual efficiency loss is given in

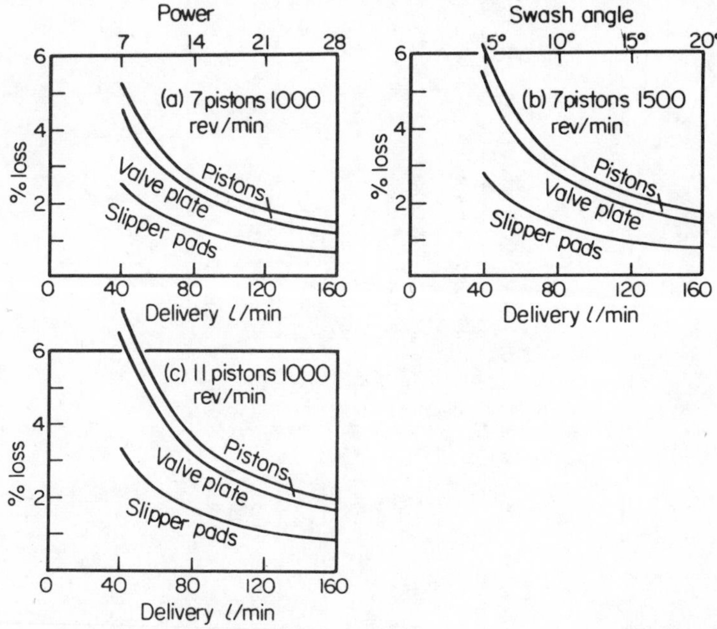

Figure 5.25. Effect of speed and number of pistons on the variation of slipper, valve plate and piston losses with delivery of 160 litre/min, 70 bar pump

terms of a loss parameter K (see *Figure 5.24b*) whose value is a function of the number of pistons, m_p. An interesting comparison is provided by the results shown in *Figure 5.25* which illustrates the distribution of the losses of three machines designed to operate at a pressure of 70 bars, and a maximum flow rate of 160 litre/min with minimum loss conditions for the slipper pads, pistons and valve plate. It can be seen from the figure and *Table 5.2* that low speeds and a small number of large pistons give better results than high speeds and a larger number of smaller pistons.

Table 5.2 DIMENSIONS OF A 160 LITRE/MIN 70 BAR IN PUMP SELECTED FOR MINIMUM VALUES OF SLIPPER, PISTON AND VALVE PLATE LOSSES

Values chosen				*Values determined for minimum losses*							
Speed rev/min	*Number of pistons*	*Slipper film gap mm × 10²*	*Piston radial clearance mm × 10²*	*Relative piston length*	*Relative piston diameter*	*Relative piston circle radius*	*Relative inner pad diameter*	*Relative port width*	*Relative valve plate clearance*	*Full delivery loss %*	*Quarter delivery loss %*
N	m_p	h	t	L	D_s	R	D_1	w	c	—	—
1 000	7	1.25	1.25	0.55	1.23	1.83	1.24	0.49	0.48	1.4	5.2
1 500	7	1.25	1.25	0.47	1.08	1.66	1.15	0.43	0.51	1.66	6.1
1 000	11	1.25	1.25	0.415	0.91	2.12	0.88	0.36	0.5	1.94	7.1

5.7.5 Axial and tilting stiffnesses of slipper pads

It is essential when designing slipper pad bearings for use in axial piston machines to ensure that they have adequate axial stiffness so that the inertia force associated with their reciprocating motion does not produce solid contact between the slipper face and the swashplate. The stiffness, s, of the slipper bearing shown is

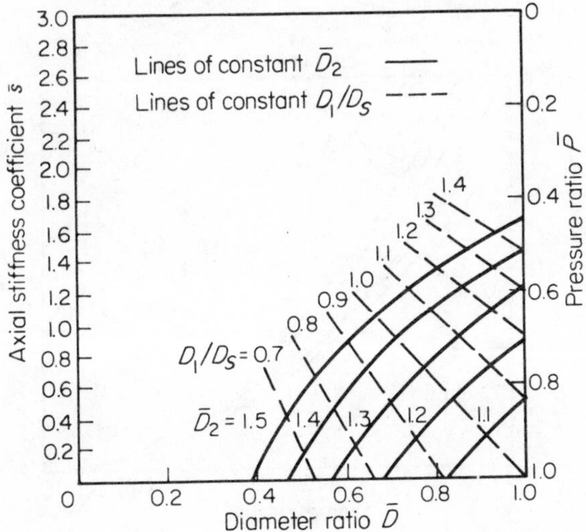

Figure 5.26(a). Variation of axial stiffness coefficient with diameter ratio for zero swash angle

obtained by deriving the value of P_r, the recess pressure in terms of the operating pressure from equations 5.18b and 5.18c and substituting this value in equation 5.19a for the vertical force F_p acting on the piston giving

$$F_p = \frac{\pi}{8} \frac{P r_1^4 D_2^2 (1-\bar{D}^2)\cos\theta}{\{\frac{4}{3} L_1 h^3 + r_1^4 \log(1/\bar{D})\}} \tag{5.25}$$

Differentiating this equation for F_p with respect to h gives the stiffness s and it is convenient to consider an axial stiffness coefficient $\bar{s}$ given by

$$\bar{s} = s/(-F/h) = 3\left\{1 + \frac{2\log\bar{D}}{\bar{D}_2^2(1-\bar{D}^2)\cos\theta}\right\} \tag{5.26a}$$

$$= 3(1-\bar{p}_r) \tag{5.26b}$$

where $\bar{p}_r = P_r/P$.

The variation of this axial stiffness coefficient with the inner and outer slipper pad diameters D_1 and D_2 is shown in *Figure 5.26a* (see Reference 38). It can be seen that if either of these diameters is altered then the value of the recess pressure, P_r, required for force balance changes (as shown by the right hand side scale of the figure) and the film thickness, h, can only be kept constant if the choke size, expressed as r_1^4/L_1, is adjusted. The values of $\bar{D}$ at which the

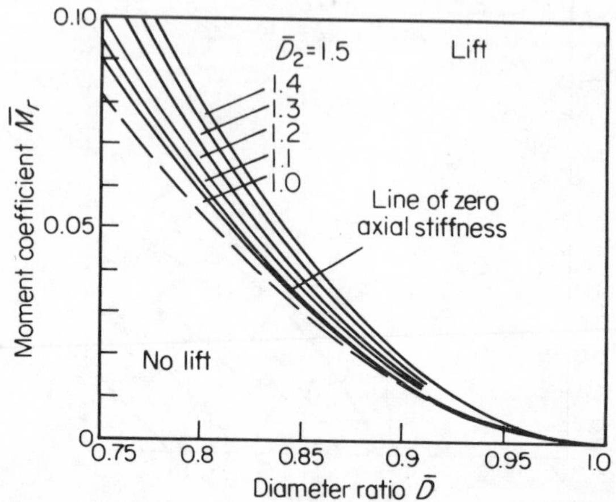

Figure 5.26(b). Derived variation of moment coefficient with diameter ratio

stiffness coefficient is zero represent design limits since the pressure in the recess is then equal to the operating pressure and this corresponds to the condition when no choke is present.

Fisher[39] has shown that the righting moment, M_r, produced when a slipper pad is tilted such that its bearing face is no longer parallel to the swashplate surface is given by

$$M_r = \pi P_r D_2^3\{f_1(\bar{D})\varepsilon_p + f_2(\bar{D})\varepsilon_p^3 + \cdots\} \tag{5.27a}$$

where ε_p is the ratio of the angle of tilt to that angle [equal to $\tan^{-1}(2h/\bar{D}_2)$] at which contact between the edge of the pad and the swashplate would occur. The righting moment is due to a redistribution of the pressure in the fluid film between the slipper pad sill and the surface of the swashplate and opposes the tilting action of any forces (e.g., centrifugal force acting at the centre of gravity of the pad) which tend to tilt the slipper pad from the position at which its face is parallel to that of the swashplate.

A tilting stiffness coefficient $\overline{M}_r$ may be defined by

$$\overline{M}_r = \left.\frac{\partial M_r}{\partial \varepsilon_p}\right|_{\varepsilon_p=0} \times \left(\frac{128 \cos \theta}{3\pi P D_s^3}\right)$$

and since $f_1(\bar{D}) = 3\{(1+\bar{D}^2)(1-\bar{D}^2)/\log(1/\bar{D}) - 4\bar{D}^2\}/256$ and P_r is given in terms of P by equation 5.19b, the tilting stiffness coefficient at $\varepsilon_p = 0$, may then be written as:

$$\overline{M}_r = \bar{D}\left\{(1+\bar{D}^2) + \frac{4\bar{D}^2 \log(\bar{D})}{(1-\bar{D}^2)}\right\} \tag{5.27b}$$

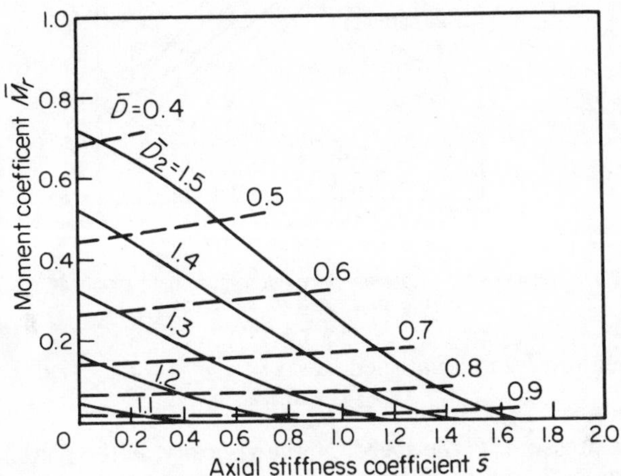

Figure 5.26(c). Variation of axial stiffness coefficient with moment coefficient for various values of $\bar{D}_2$

The variation of $\overline{M}_r$ with the diameter ratio $\bar{D}$ has been derived in Reference 38 (from which *Figure 5.26b* is taken) and as $\bar{D} \to 1.0$ the value of $\overline{M}_r$ is given by $\overline{M}_r \to 4\bar{D}_2(1-\bar{D})^2/3$. The line of zero axial stiffness, equation 5.26a with $\bar{s} = 0$ is also shown in *Figure 5.26b* and it should be noted that, unlike the axial stiffness coefficient, the tilting stiffness coefficient decreases with increasing values of D_1. It does, however, increase slightly with D_2 and these conflicting requirements for large values of both stiffnesses are best shown in *Figure 5.26c*. Further work on this topic for dynamic conditions has been presented by Royle and Raizada[40] where pressure contours beneath the pad have been derived.

It has been shown in the previous section that conditions of minimum power loss occur for a given value of film thickness, h,

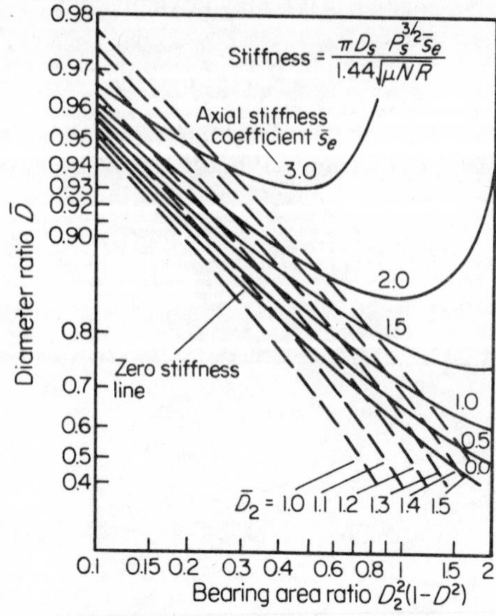

Figure 5.27. Variation of pad geometry and axial stiffness coefficient with pad bearing area ratio (best efficiency conditions)

when equation 5.22b is satisfied, i.e.

$$\beta = 0.129/\bar{h}^2 \tag{5.22b}$$

and if the pressure, P, the speed, N, the viscosity, μ, the piston pitch circle radius, R, and the piston diameter D_s, are previously selected then this may be written as

$$h = k_e\sqrt{\{\bar{D}_2^2(1-\bar{D}^2)\}}$$

where $k_e = D_s\sqrt{\{0.129\mu N\bar{R}/P\}}$.

Substituting this value of h into equation 5.26a and defining a minimum loss axial stiffness coefficient, $\bar{s}_e$, given by $\bar{s}_e = -k_e s/F$ then gives

$$\bar{s}_e = \frac{3}{\{\bar{D}_2^2(1-\bar{D}^2)\}^{1/2}}\left\{1+\frac{2\log(\bar{D})}{\bar{D}_2^2(1-\bar{D}^2)\cos\theta}\right\} \tag{5.28}$$

with the actual stiffness, s, given by

$$s = \frac{\pi D_s p^{3/2}}{1.44(\mu N\bar{R})^{1/2}}\bar{s}_e$$

It will be recalled that in section 5.7.2 it was noted that the required value of β_s for minimum power loss could be obtained for the conditions of a fixed value of film thickness, h, by varying D_1, D_2 or both such that the necessary value of the ratio of the areas of the pad sill and piston $\bar{D}_2^2(1-\bar{D}^2)$, was obtained and gave

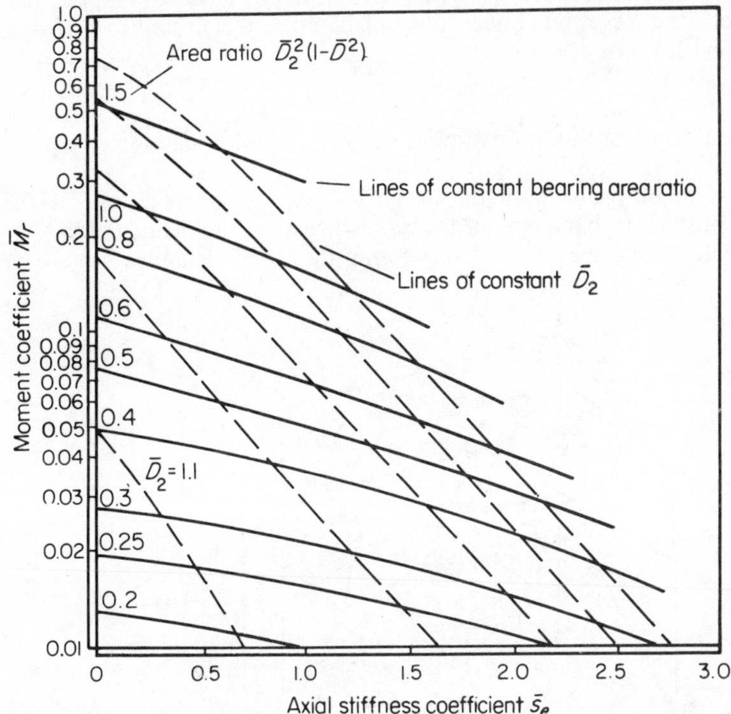

Figure 5.28. Variation of moment coefficient with axial stiffness coefficient for given values of bearing area ratio (best efficiency conditions)

the required value of β_s. As a result any combination of values of D_1 and D_2 may be used to satisfy this requirement and *Figure 5.27* shows the variation of $\bar{D}$ with bearing area ratio for various values of $\bar{D}_2$. Since the values of both $\bar{D}$ and $\bar{D}_2^2(1-\bar{D}^2)$ are fixed at any point on this figure the value of $\bar{s}_e$ is also fixed (equation 5.28) and curves of constant $\bar{s}_e$ are also shown.

The tilting moment coefficient is unaffected by variations in film thickness and in general terms it can be said to have no alternative form if the conditions of minimum loss exist. Its variation with the diameter ratio, $\bar{D}$, and bearing area ratio can however be shown in the same manner as the axial stiffness results[38] given in *Figure 5.27*.

If a pad is to be designed to give the lowest power loss for a specified value of film thickness, h, and this demands a given value of bearing area ratio to satisfy equation 5.22b then *Figure 5.28* may be used to select values of the inner and outer diameters of the slipper pad such that acceptable values of both axial and tilting stiffnesses are obtained.

Further work on the general subject of the design slipper pad bearings operating under dynamic conditions is described by Royle and Raizada.[40]

5.7.6 Axial thrust balancing in axial piston machines

A major problem in the design of axial piston machine is the satisfactory balancing of the axial thrust on the cylinder block or rotor (see *Figure 5.18*). Considering the small clearance *c* between

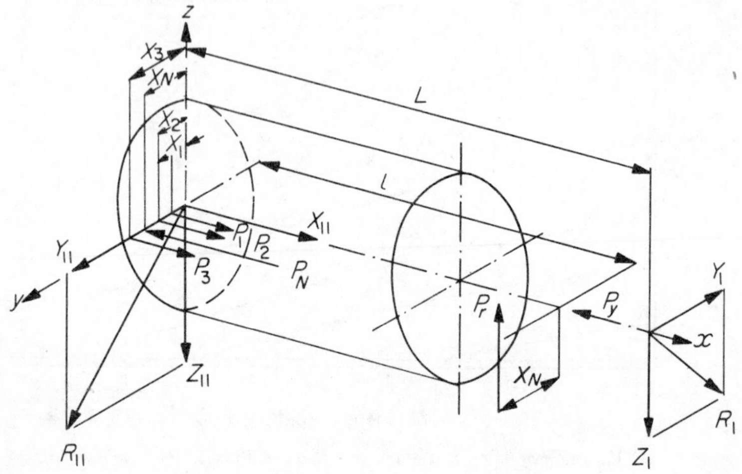

Figure 5.29. System studied by Saitchenko

the adjacent faces of the cylinder block or rotor and the valve plate of the machine as shown in *Figure 5.24a*, it can be seen that if the closing or hydrostatic force associated with the high pressure pistons and cylinders is excessive then the faces may touch and wear will occur.

Alternatively, if the opening or hydrostatic valve plate force associated with the pressure in the high pressure kidney port and across its lands is too great then the rotor will be forced away from the valve plate and excessive leakage losses will occur. Both the piston and valve plate forces generally vary when the shaft of the machine is rotating due to different numbers of cylinders being pressurised and to variations in the effective angular length of the high pressure, valve plate kidney port produced by its effective extension by the cylinder block ports as shown in *Figure 5.18b*. The importance of obtaining a satisfactory means of balancing these forces is illustrated by the numerous inventions in this field, some of which are reviewed by Shute and Turnbull[41] and the designs by Bowers and Thoma are of particular interest since these are known to have been fully developed.

Saitchenko[42] has derived a simplified analysis of the force system shown in *Figure 5.29* and suggests that the high pressure piston force should exceed that produced by the pressure distribution in the valve port and across its lands by at least 15% and that the moment of the piston force about the valve plate's TDC and BDC axis should also

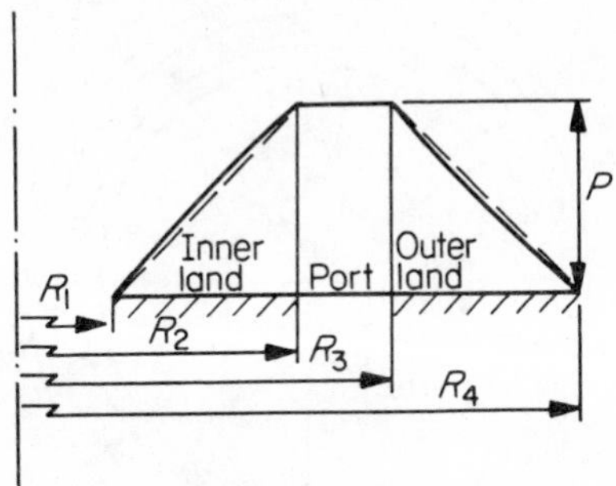

Figure 5.30(a). Typical pressure distribution across valve plate port and lands $R_2/R_1 = 1.2$; $R_3/R_1 = 1.3$; $R_4/R_1 = 1.5$ *(--- linear approximation)*

exceed that of the valve plate force by at least 15%. Guillon[43], however, suggests that adequate force balance exists if the valve ports are considered to extend 180° ('end effects', that is pressure between the inlet and outlet ports, may then be ignored) and the valve plate force is between 5% and 15% less than the average piston force which is in agreement with the figure quoted by Saitchenko[42].

Work by Ernst[45] is of a semi-empirical nature and based on the use of a 'pressure ratio factor', k_p, which is apparently used to account for the non-linear pressure drop across the valve plate lands (and the frictional force between the pistons and their cylinders). His approach can be followed by reference to *Figure 5.30a*, which shows the hydrostatic pressure distribution across the high pressure ports and lands of a valve plate. Including the effect of the area of the cylinder ports in both expressions for the forces and referring to *Figure 5.30b*, shows that the axial force due to the pressure distribution across the valve plate port and lands is given by:

$$\frac{\beta_v}{2}\left\{(R_3^2-R_2^2)P+(R_4^2-R_3^2)\frac{P}{k_p}+(R_2^2-R_1^2)\frac{P}{k_p}\right\}$$

where k_p is a pressure ratio factor introduced by Ernst and the notation is as shown in *Figure 5.30b*.

The average piston force is given by $\pi D_s^2 P m_p/8$ where m_p is the total number of pistons and equating this to the valve plate force above with $\beta_v = \pi$ gives

$$(R_3^2 - R_2^2)(k_p - 1) + (R_4^2 - R_1^2) = D_s^2, m_p k_p/4 \tag{5.29}$$

which is the formula quoted by Ernst. The assumption that $\beta_v = \pi$ is probably made to account for the decaying pressure distribution

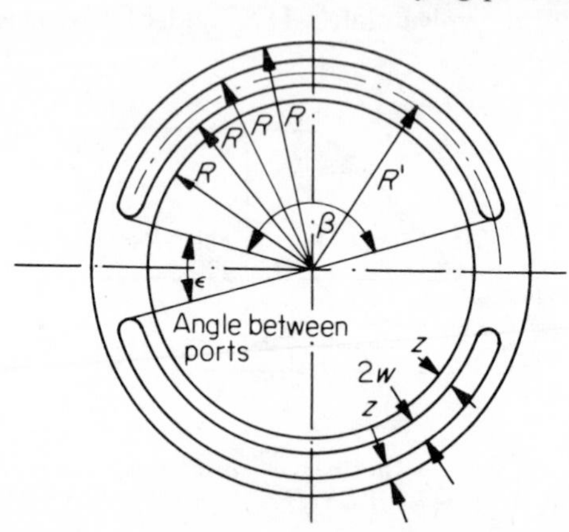

Figure 5.30(b). Valve plate geometry. (Port width ratio $w' = w/(w+z)$; Bearing width ratio $\bar{b} = 2(w+z)/R'$)

between the ends of the high and low pressure kidney ports but more recent work by Shute and Turnbull[46] has shown that in a typical configuration the effective angular port length for the valve plate is approximately 194° for half of the time and approximately 154° for the other half so that its mean effective length is 174°. Whilst this suggests that if a mean value of 180° is taken the error is only 3%, it should be noted that this is in addition to any error produced by suggesting $k_p \neq 2$ which would imply that a significant allowance must be made for the effects of the curvature of the lands of the valve plate. This is not generally the case and the actual effect of the curvature of the valve plate lands on the pressure drop across them is usually negligible (it tends to be self-cancelling) as is shown in *Figure 5.30a* for a typical valve plate geometry.

In addition to describing a method of deriving the effective fluctuations of the hydrostatic force for the valve plate by an analogue technique, based on the use of electrically conducting paper (see *Figure 5.31*), Shute and Turnbull[46] also give results in a polar

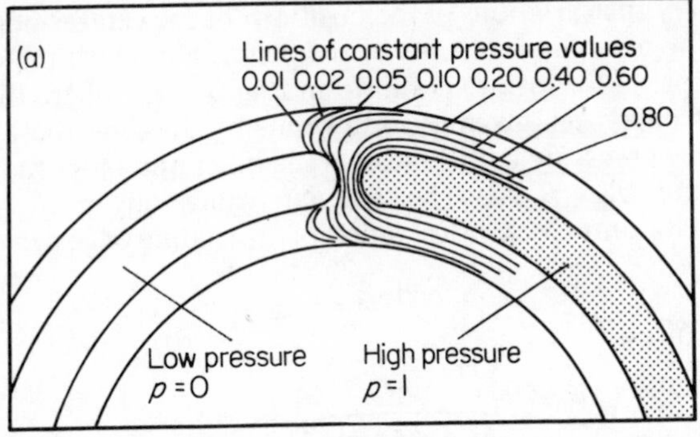

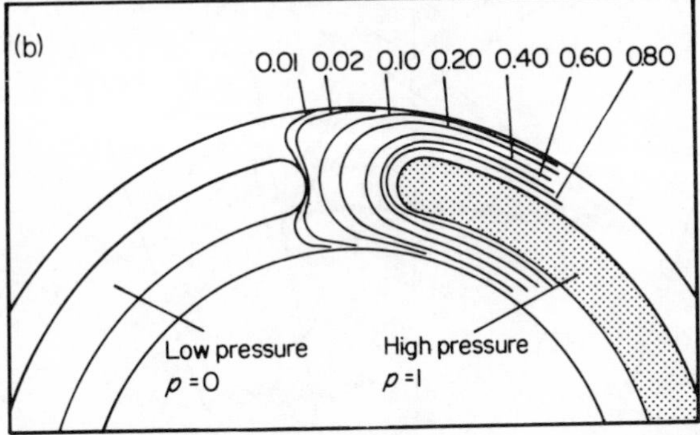

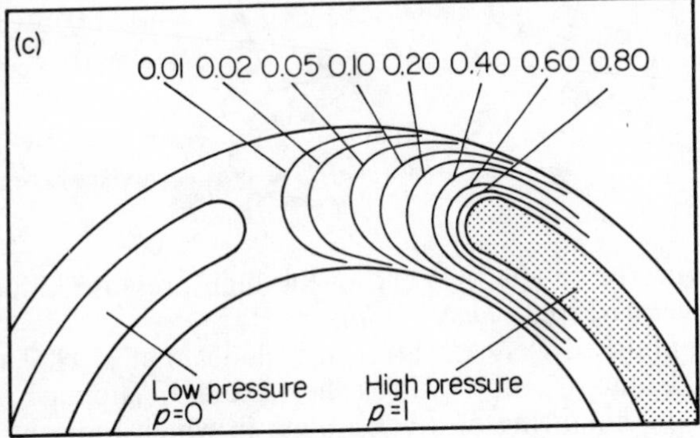

Figure 5.31. Derived variation of the pressure distribution around the ends of valve plate ports with the angle between the ports ($\bar{b} = 0.4$; $w' = 0.4$)
(a) Angle between ports = 4° (b) Angle between ports = 16°
(c) Angle between ports = 40°

diagram for the variations of the positions of the centres of pressure and for the resultant moments of the piston and valve plate forces. An example of the former is shown in *Figure 5.32*, where the radial distances have been given as radii ratios by dividing the radius to the centre of pressure by the mean of the inner and outer radii of the valve plate $(R_1+R_4)/2$. The angular values given in brackets represent the anti-clockwise rotation of the leading edge of a cylinder

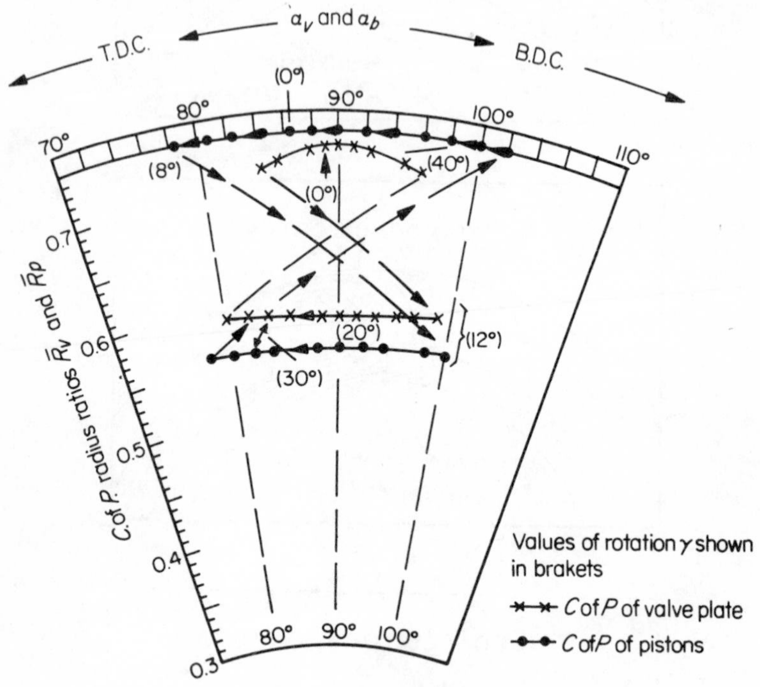

Figure 5.32. Variation of positions of the C. of P. of the valve plate and piston forces

port beyond the geometrical end of the high pressure kidney port of the valve plate (see *Figure 5.18b*).

Other published work in this field includes that of H. Thoma[47], Nation[48], Franco[49], Blok[50], Prokofiev[51] and J. Thoma[52]. Nation describes the balancing of a valve plate in which the central area was pressurised and as a result the work is not directly applicable to the usual machine configuration. Franco includes the effect of centrifugal force on the fluid in the film in his analysis and this is later shown to be negligible for the usual range of operating conditions, whilst Blok has shown how the operating conditions and some geometrical features can affect the performance of a transmission. Further details of J. Thoma's work are given in section 5.7.7.

An interesting suggestion has been made by Palmer *et al*[53] concerning the use of small hydrostatic bearings situated in the valve plate lands as shown in *Figure 5.33a.* Stiffnesses of the order of 10^6 N/mm, can be achieved with such a device but it should be remembered that displacements of the order of only 10^{-3} mm are likely to be encountered. A more practical value of stiffness is therefore usually considered, namely, the change of force produced by a change of film thickness of 10^{-3} mm. Even so, a movement of 10^{-3} mm can result in a force change of several hundred newtons which represents a useful contribution to the rigidity of the system. Palmer[53] *et al* has shown that by careful design, a valve plate configuration can be derived which not only possesses a considerable stiffness and righting moment but which also compares favourably with the plain design from the point of view of minimum efficiency loss as is shown in *Figures 5.33b*, *5.33c* and *5.33d*.

5.7.7 Effects of variable viscosity

The viscosity of most hydraulic fluids falls rapidly with increasing temperature and rises with increasing pressure so that in any system where either or both of these quantities change, some account should be taken of their effect on the viscosity of the operating fluid.

In the previous sections of this chapter the value of the viscosity in the equations has been assumed to be equal to a 'mean effective' value and over a limited range of operating conditions (e.g. see *Figure 5.21b*) this form of approximation yields satisfactory results. However, in view of the sometimes very wide range of conditions under which many modern systems are required to operate the measurement or even the choice of a satisfactory value of the mean effective viscosity is very difficult and a more critical examination of the effects of variations in the fluid viscosity is often desirable. This may be done if the viscosity is written in terms of the instantaneous values of the fluid pressure, P, and temperature, T, and a convenient form is

$$\mu = \mu_0 \, e^{(\alpha P - \beta T)}$$

where α and β are the pressure and temperature coefficients of viscosity.

As a simple example, the poiseulle flow along a small diameter tube may be examined and initially it will be assumed that the work

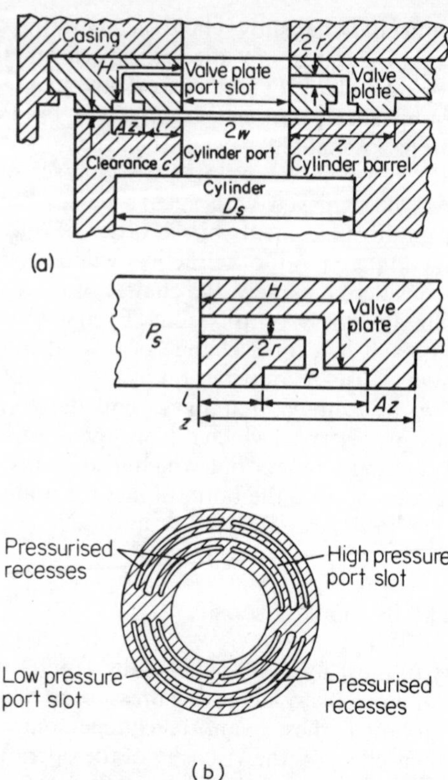

Figure 5.33. The hydrostatic valve plate
(a) General arrangement of hydrostatic valve plate
(b) Valve plate having stiffness and inherent righting moment

done in forcing the fluid through the tube is entirely absorbed by the fluid, thereby raising its temperature. In other words no heat will be lost through the wall of the tube. Referring to *Figure 5.34*, the flow rate q is given by

$$q = -\frac{\delta p}{\delta x}\frac{\pi r_1^4}{8\mu} \tag{5.30}$$

and over a small length δx at a distance x from the high pressure end the pressure gradient may be related to the temperature gradient by considering the energy balance giving

$$-q\frac{\delta p}{\delta x} = \rho c q\frac{\delta T}{\delta x} \tag{5.31a}$$

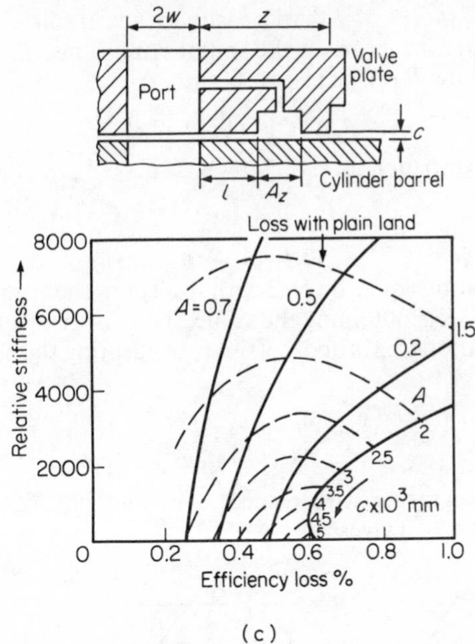

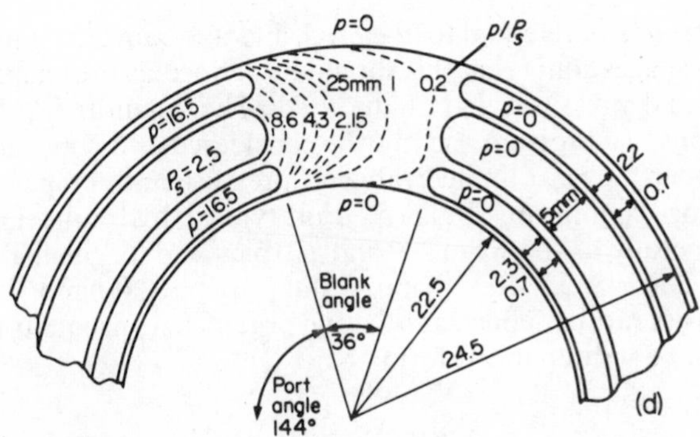

Figure 5.33. (continued)

(c) *Variation of stiffness with efficiency loss for different values of recess/total land width ratio (lines of constant c – – – –)*

(d) *Pressure contours on hydrostatic valve plate ($\bar{p} = 0.65$) (obtained by conducting paper technique)*

where ρ is the fluid density and c is its specific heat containing the mechanical equivalent of heat term. Integrating equation 5.31a with respect to x and inserting the limits $P = P_1$ when $T = T_1$ then gives

$$P = P_1 + \rho c(T_1 - T) \tag{5.31b}$$

and since $\mu = \mu_o e^{(\alpha P - \beta T)}$ the viscosity at any point along the tube may be written in terms of the initial conditions, P_1 and T_1, and the temperature T only, i.e.

$$\mu = \mu_o e^{\alpha(P_1 + \rho c T_1) - (\beta + \rho c \alpha)T}$$

which may be written as

$$\mu = E e^{KT} \tag{5.32}$$

where $E = \mu_o e^{\alpha(P_1 + \rho c T_1)}$ and $K_1 = -(\beta + \rho c \alpha)$.

Differentiating equation 5.32 with respect to the temperature T to give $\delta\mu/\delta T = K\mu$, obtaining the value of δT in terms of δp and then δx from equations 5.31s and 5.30 and substituting this into the result gives

$$\frac{\delta\mu}{\delta x} = \mu^2 \frac{8Kq}{\rho c \pi r_1^4}$$

which on integrating with respect to x and putting in the limits $\mu = \mu_1$ when $x = 0$ gives

$$\mu = \frac{\mu_1}{1 - \left\{\dfrac{K}{\rho c}\right\} \dfrac{8q\mu_1 x}{\pi r_1^4}}$$

The term $-(K/\rho c)$ is equal to $(\alpha + \beta/\rho c)$ and it is convenient to follow Thoma's suggestion[52] that this should be defined as the reciprocal of the 'thermal pressure', P_T, of the fluid. The quantity P_T has the dimensions of a pressure and may be considered as a constant and a fluid property over a limited range of pressure and temperature as shown for example by Fuller.[54] For typical hydraulic fluids its value generally lies between 200 and 270 bars and for Shell Tellus 27 hydraulic oil $\alpha \approx 2.2 \cdot 10^{-3}(\text{bars})^{-1}$ and $(\beta/\rho c) \approx 2.0(\text{bars})^{-1}$. Hence the viscosity ratio $\bar{\mu}$ defined as μ/μ_1 and obtained from equation 5.33a may then be written in the form

$$\bar{\mu} = \frac{1}{1 + B_1 q\bar{x}} \tag{5.33b}$$

where $\bar{x}$ is the ratio x/L_1 and B_1 is the reciprocal of the rate of flow of a fluid having a constant viscosity μ_1 through a tube of length L_1 produced by a pressure drop of P_T. The value of this flow rate is given by

$$\frac{1}{B_1} = \frac{P_T \pi r_1^4}{8\mu_1 L_1} \tag{5.34}$$

The variation of the pressure (expressed as a fraction of P_T) along the tube may be obtained by substituting the value of μ ($\mu = \bar{\mu}\mu_1$), from equation 5.33b into equation 5.30 and integrating with respect to x with the limit $(p/P_T) = 0$ when $\bar{x} = 1$, i.e.

$$\frac{p}{P_T} = \bar{P} = \log\left\{\frac{1+B_1 q}{1+B_1 q\bar{x}}\right\} \tag{5.35a}$$

and the actual flow rate, q, can be obtained from this equation since $p = P_1$ or $\bar{P} = \bar{P}_1$ (where $\bar{P}_1 = P_1/P_T$) when $\bar{x} = 0$, giving

$$\bar{P}_1 = \log(1+B_1 q) \tag{5.35b}$$

or

$$q = \frac{1}{B_1}\{e^{\bar{P}_1} - 1\}$$

$$= \frac{P_T \pi r_1^4}{8\mu_1 L_1}\{e^{\bar{P}_1} - 1\} \tag{5.35c}$$

If the pressure drop, P_1, along the tube is small compared with the thermal pressure, P_T, then the index term in the above equation is small and the value of q approaches the value q_1 which is obtained if the viscosity had a constant value μ_1. This is given by

$$q_1 = \frac{P_1 \pi r^4}{8\mu_1 L} \tag{5.35d}$$

Some measure of the error introduced by neglecting the effect of pressure and temperature variations particularly at high pressures can be obtained by dividing the actual flow rate equation 5.35c by its approximate, or low pressure form, 5.35d given above and the result in the form of a flow rate ratio—$v-\bar{P}_1$ curve is shown in *Figure 5.34b*.

It is of interest to note that the above method of approach has also been applied to the analysis of the flow along a thin gap when one of the boundaries is moving relative to the other. In his study of such a system Wilson,[55] assumed that the viscosity pressure coefficient, α, was zero and considered the condition with the velocity of the moving boundary in the same plane as the pressure gradient and parallel to the length of the gap.

De Raucourt[56] after showing that the 'thermal wedge' effect could probably be neglected in comparison with thermal distortion effects (see below) when considering plane thrust bearings, considered the system with the velocity of one boundary perpendicular to the plane of the pressure gradient and he also assumed the viscosity pressure coefficient, α, was zero. Both he and Wilson, however, allowed for the additional rise in temperature produced by the external work done in moving one of the boundaries, i.e., in shearing

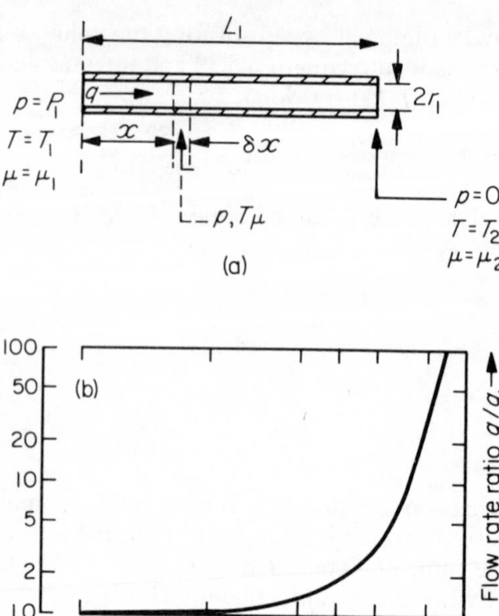

Figure 5.34. Flow along tube when effects of the viscosity coefficients are included in analysis $\mu = \mu_0 \cdot e^{\alpha p - \beta T}$
(*a*) *geometry and end conditions*
(*b*) *Variation of flow rate/flow rate with* $\alpha = \beta = 0$ *with the inlet pressure ratio* $\bar{P}_1$

the fluid film, and other aspects of the problem have been considered by Abir *et al.*[57] Thoma, [52] however, suggested that this term could be neglected, thereby making an arbitrary allowance for the heat lost by conduction to the boundaries but he did allow for the effect of both pressure and temperature on the viscosity of the fluid by introducing the viscosity pressure and temperature indices. Both de Raucourt and Thoma were considering a system corresponding to that of the fluid film between the adjacent faces of a cylinder block and valve plate in an attempt to improve the force balance conditions (see section 5.7.6) as were the authors of References 58 and 59.

These last two references give an analysis in which neither the viscosity pressure coefficient nor the additional temperature rise caused by the work done in moving one of the boundaries were neglected and the resulting variation of the flow rate with operating pressure for a range of values of boundary velocities is shown in *Figure 5.35b* together with some typical pressure profiles across the

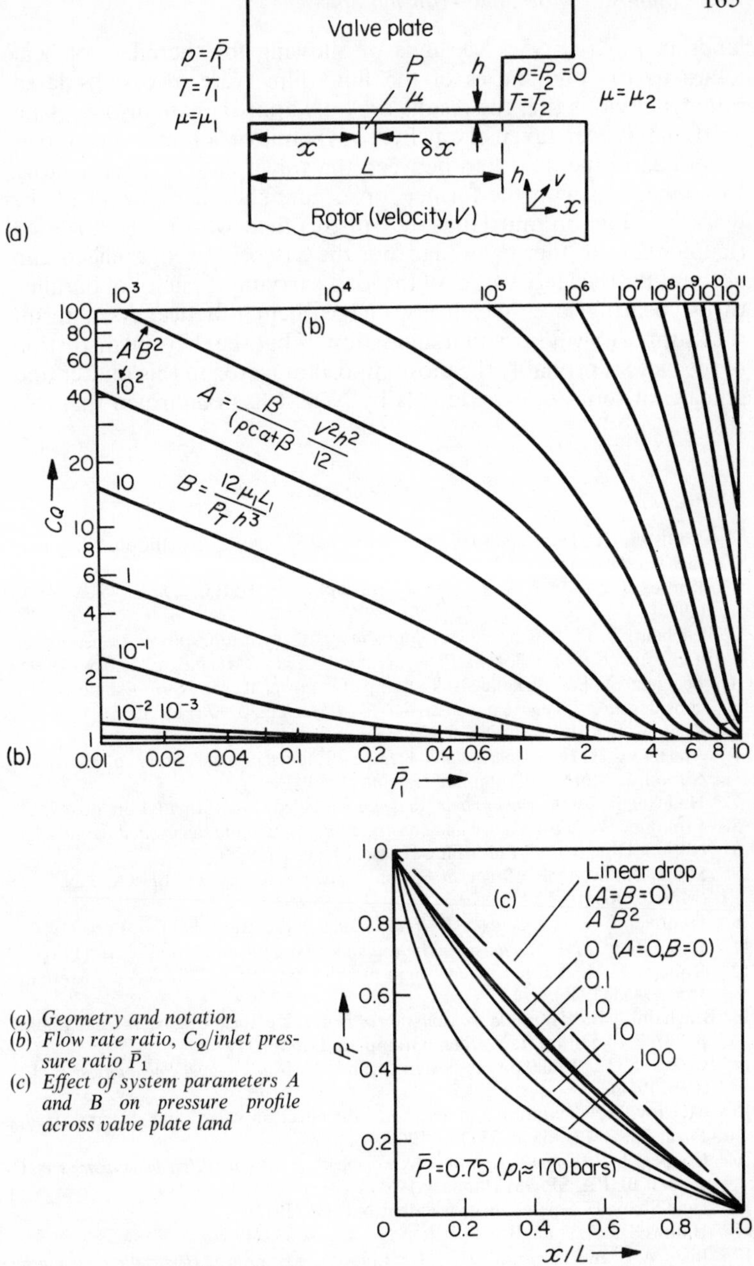

(a) Geometry and notation
(b) Flow rate ratio, C_Q/inlet pressure ratio $\bar{P}_1$
(c) Effect of system parameters A and B on pressure profile across valve plate land

Figure 5.35. Fluid film conditions with relative velocity between rotor and valve plate faces

lands in *Figure 5.35c*. Methods of allowing for the effect of heat losses to the boundaries of the fluid film were also considered but it is obvious that the results are only approximate insofar as no account has been taken of any hydrodynamic or 'squeeze film' forces generated in the fluid film between the valve plate and rotor faces. In connection with the former force, note should be taken of the work by Dowson and Hudson[60] in this field who have suggested that neither the 'thermal wedge' nor the 'viscosity wedge' effects can account for the large values of the load carrying capacity of parallel, thrust bearing faces which are met with in practice. As a result they support Swift's original suggestion[61] that the thermal distortion of the faces is probably the most important factor in this matter and an elegant series of experiments by Neal[62] has confirmed that this is so.

REFERENCES

1. Addison, H., *Centrifugal and other rotodynamic pumps*, 3rd edition, Chapman and Hall, London (1965).
2. Rouse, H. and Ince, S., *History of hydraulics*, 1st edition, Dover, New York (1963).
3. Beacham, T. E., 'Historical development of oil hydraulic power transmission and control' *Proc. Oil, Hyd. Power Trans. & Cont., I.Mech.E.*, p. 3 (Nov. 1961).
4. Beacham, T. E., 'Variable-stroke pumps for power transmission—some design considerations'. *Proc. Conf. Hyd. Servos., I.Mech.E.*, p. 17 (Feb. 1953).
5. Rigg, A., British Patent No. 6047 (1885).
6. Anderson, H. H., *Liquid pumps*, Kempe's Engineers Year Book, p. 833, 75th edition, **1,** Morgan–Grampian, London (1970).
7. Hadekel, R. *Displacement pumps and motors*, 1st edition, Pitman, London (1951).
8. Prokofiev, V., *Basic theory and construction of hydraulic pumps and motors*, 1st edition, High School Publishing House, Moscow (1968).
9. Shute, N. A. *et al.*, *Survey of hydraulic pumps and motors*, BHRA, T.N. 765 (March 1963).
10. Himmler, C. R., *La Commande Hydraulique*, 2nd edition, Dunod, Paris (1960).
11. Faisandier, J. *Les Mécanismes Hydrauliques*, 1st edition, Dunod, Paris (1957).
12. Roberts, F. H., 'Large capacity, high pressure pumping plant'. *Proc. I.Mech.E.* **155,** p. 453 (1946).
13. Beacham, T. E., *Hydraulic transmission of power*, Kempe's Engineers Year Book, p. 797, 75th edition, **1.** Morgan–Grampian, London (1970).
14. Cozza, S. L., 'Fluid power design forum', *Hydraulics and Pneumatics*, p. 12 (Feb. 1966).
15. McCloy, D., 'Cavitation and aeration—the effect on valves and systems'. *Hyd. Pneu. Power* **12,** 133, p. 33 (Jan. 1966).
16. Cooper, J. E., *Some design and development aspects of tilting head pumps and motors*, BHRA, SP. 981 (January 1969).
17. Hele-Shaw, H. S. *et al.*, British Patent No. 336 (1912).

17a. Halliwell, J. A. *et al.*, British Patent No. 1, 158, 638 (1969).

17b. Read, A. C. and Turnbull, D. E., 'The Epicyclic gear pump', *Hydraulic Pneumatic Power* (Nov. 1961).

18. Shute, N. A. and Turnbull, D. E., 'Minimum power loss of hydrostatic slipper

bearings for axial piston machines', *Proc. Conf. Lub. Wear., I.Mech.E., Paper No. 1*, p. 6 (May 1963).

19. Shute, N. A. and Turnbull, D. E., 'Minimum power loss conditions of the pistons and valve plate in axial-type pumps and motors', *ASME Paper No. 63-WA-90* (Nov. 1963).
20. Beacham, T. E., 'Gear pumps', *Proc. I.Mech.E.* **155,** p. 417 (1946).
21. Mortenson, P. C., 'Trends in hydraulic equipment for mobile construction machinery', *Proc. Conv. Applic. of Hyd., I.Mech.E.*, p. 9 (Oct. 1963).
22. Grosser, C. E., 'Controlling pulsations in hydraulic equipment'. *Machine Design* (April 1941).
23. Schlösser, W. M. J., 'A mathematical model for displacement pumps and motors'. *Hydraulic Power Transmission*, Nos. 4, p. 252 and 5, p. 324 (1961).
24. Wilson, W. E., *Positive displacement pumps and fluid motors*, 1st edition, Pitman, London (1950).
25. Schlösser, W. M. J., *The overall efficiency of positive-displacement pumps*, BHRA, SP. 983 (January 1969).
26. Platt, A. and Kelly, E. S., *Life testing of hydraulic pumps and motors on fire resistant fluids*, BHRA, SP. 982 (January 1969).
27. Price, C. K. J., 'Control of rotary power transmission', *Proc. Conf. Oil Hyd. Power, Trans. & Cont., I.Mech.E.*, p. 41 (Nov. 1961).
28. Rigby, R. W., *An integral control constant pressure device with built-in stabilisation for a variable delivery axial piston hydraulic pump*, BHRA, SP. 988 (January 1969).
29. Bowers, E. H. 'Hydrostatic power transmission for vehicles', *Proc. Conf. Oil Hyd. Power, Trans. & Cont., I.Mech.E.*, p. 41 (Nov. 1961).
30. Westbury, R. *et al.*, 'A double-differential, hydrostatic, constant-speed, alternator drive', *Proc. Conf. Oil Hyd. Power, Trans. & Cont., I.Mech.E.*, p. 50 (Nov. 1961).
31. Constantinesco, A., 'A new approach to fluid power transmission analysis', *Proc. Conf. Oil Hyd. Power, Trans. & Cont., I.Mech.E.*, p. 69 (Nov. 1961).
32. Petty, M. G. R., 'Development in hydraulic transmission drives for ship's cargo winches', *Conf. Hyd. Mech., I.Mech.E.*, p. 9 (March 1954).
33. Beacham, T. E., 'Positive displacement machinery for power transmission', *Proc. Conf. Hyd. Mech., I.Mech.E.*, p. 33 (March 1954).
34. Edgehill, C. M., *Some factors determining the choice of a particular hydrostatic transmission unit*, BHRA, SP. 985 (January 1969).
35. Blackburn, J. E. *et al.*, *Fluid power control*, 1st edition, M.I.T., Cambridge, Mass. (1960).
36. Chesnut, H. and Mayer, R. W., *Servomechanisms and regulating system design*, 2nd edition, **1,** Wiley, New York (1959).
37. McCallion, H., Dudley, B. R., and Knight, G. C., *Analysis of a dynamically loaded hydrostatic transmission system*, BHRA, SP. 986 (January 1969).
38. Shute, N. A. and Turnbull, D. E., *The axial and tilting stiffnesses of hydrostatic slipper bearings*, BHRA, R.R., 759 (March 1963).
39. Fisher, M. J., *A theoretical determination of some characteristics of a tilted hydrostatic slipper bearing*, BHRA, R.R. 728 (April 1962).
40. Royle, J. K. and Raizada, R. S., 'Numerical analysis of effects of tilt, sliding and squeeze action on externally pressurised, oil-film bearings'. *Proc. Conv. Lub. & Wear, I.Mech.E.* (Holland) (May 1966).
41. Shute, N. A. and Turnbull, D. E., *A review of some recent developments in the design of axial piston machines*, BHRA, T.N. 793 (Jan. 1964).
42. Saitchenko, J. S., 'Force balance conditions of the valve plate and rotor of an axial piston pump', *Stankii Instrument*, **10,** p. 28/9 (1950). Translated by BHRA, T. 750 (March 1963).

43. Guillon, M., *Hydraulic servo-system analysis and design*, p. 104, Butterworths, London (1969).
44. Khaimovich, E. M., *Hydraulic control of machine tools*, 1st edition, Pergammon, London (1965).
45. Ernst, W., *Oil hydraulic power and its industrial applications*, 2nd edition, McGraw-Hill, New York (1960).
46. Shute, N. A. and Turnbull, D. E., *The thrust balancing of axial piston machines*, BHRA, R.R. 772 (May 1963).
47. Thoma, H., 'High pressure hydraulic power transmission', *Proc. I.Mech.E.* **172,** p. 29 (1958).
48. Nation, H. J., 'Some experiments on hydraulic motor distributor valve performance', *Jnl. Agri. Eng'g.* **6,** 3, p. 183 (1961).
49. Franco, N., 'Pump design by force balance', *Hydraulics and Pneumatics*, **14,** p. 101 (Nov. 1961).
50. Blok, P., 'Theoretical and experimental investigations in connection with a fluid transmission mechanism'. Zurich F.T.U., Thesis No., 2116.
51. Prokopiev, V. N. (Editor), *Variable delivery axial piston pumps, motors and transmissions*, 1st edition, Machine Construction, Moscow (1969).
52. Thoma, J., 'Sealing gaps', *Hydraulic and Pneumatic Power Control*, **9,** 105, p. 627 (Sept. 1963).
53. Palmer, K. *et al.*, 'The hydrostatic balancing of valve plates', *ASME, Paper No. 64-WA/LUB-13* (Nov-Dec. 1964).
54. Fuller, D. D., *Theory and practice of lubrication for engineers*, 1st edition, Wiley, New York (1956).
55. Wilson, W. E., 'Design of optimum clearances in positive displacement pumps and motors', *ASME Paper No. 55-5-4* (1955).
56. de Raucourt, F., 'Le Graissage de surfaces planes par "coin thermique"', *Rev. Trans. Hyd. Méch.*, **3,** p. 44 (June 1963); Translation by BHRA, T. 794 (March 1964).
57. Abir, D. *et al.*, *Topics in applied mechanics*, 1st edition, p. 179. Elsevier, Amsterdam (1965).
58. Shute, N. A. *et al.*, *The hydrostatic pressure distribution across the lands of a valve plate of an axial piston machine, Parts I, II and III*, BHRA, R. R. 795 (January 1964); R. R. 796 (March 1964) and R.R. 810 (June 1964).
59. McKeown, J. *et al.*, 'Hydrodynamic factors affecting the design of valve plates and thrust bearings', *Proc. I.Mech.E.*, 181 Pt, 1. 24 p. 653–666 (1967).
60. Dowson, D. and Hudson, J. D., 'Thermo-hydrodynamic analysis of the infinite slider-bearing. Part II: The parallel surface bearing'. Paper No. 5, *Proc. Lub. and Wear Conv., Bournemouth, I.Mech.E.* (May 1963).
61. Swift, H. W., 'Contribution to "Fluid film lubrication of parallel thrust surfaces"', *Proc. I.Mech.E.*, **155,** p. 49 (1946).
62. Neal, P. B., 'Film lubrication of plane-faced thrust pads', Paper No. 6, *Proc. Lub. and Wear Conv., Bournemouth, I.Mech.E.* (May 1963).

Chapter 6

Servo-valve Characteristics and Designs

6.1 Introduction

The basic elements of a typical hydraulic, position control servomechanism are shown in *Figure 6.1* and with very few exceptions the control valve is either a spool valve or a valve having a spool valve as its final stage. In other words, the flow rate of the working fluid to the jack or motor is generally metered by such a device.

An input signal, x, which is proportional to the required response or output position, would be fed into the system and, in the system shown in *Figure 6.1*, this signal would be a small voltage. This voltage

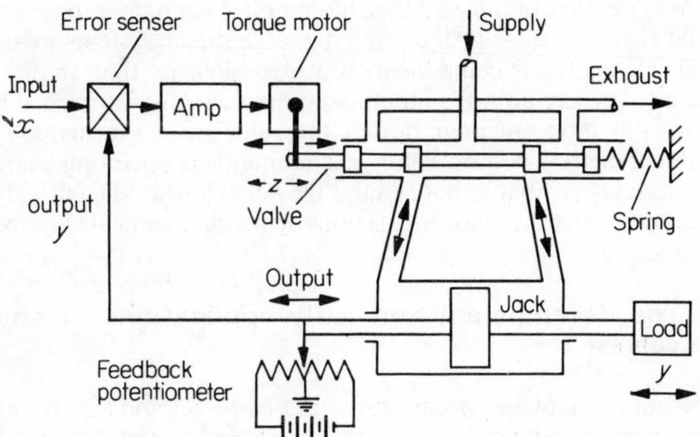

Figure 6.1. Hydraulic position control servomechanism

would pass through the error senser or differencing device and, after a voltage proportional to the output position had been subtracted from it, it would then be amplified and applied to the valve actuator or torque motor attached to the control valve. In its turn this component would open the control valve by an amount, z, proportional to the error signal and permit high pressure fluid to flow into one side of the jack and low pressure fluid to escape to exhaust or the low pressure reservoir of the system. As a result the jack would move until a voltage equal to the input voltage and proportional to

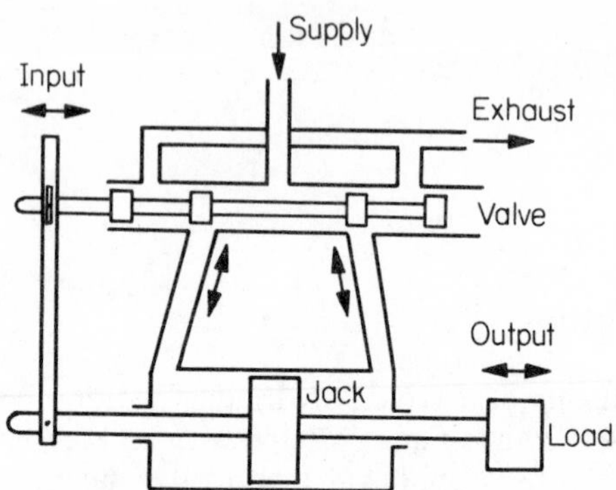

Figure 6.2. Simple valve-jack system

the jack's displacement, y, was fed back from its feedback potentiometer. The input voltage, x, at the error sensing device would then be cancelled so that at this point the signal passed on to the valve would be zero and the valve would be closed.

The system motion would then be complete and the output signal would then be equal to the input signal. Simpler systems exist in which no electrical components are used such as that shown in *Figure 6.2*, where both the input and output are displacements, but the overall mode of operation is the same as is the method of analysing their responses. Before examining their operating characteristics it is essential to understand the behaviour of the individual components and the following sections of this chapter deal with these matters.

6.2 The development and early applications of spool valves

The component of the systems shown in *Figures 6.1* and *6.2* on which most research and development effort has been expended is without doubt the control valve. A very large number of designs now exist

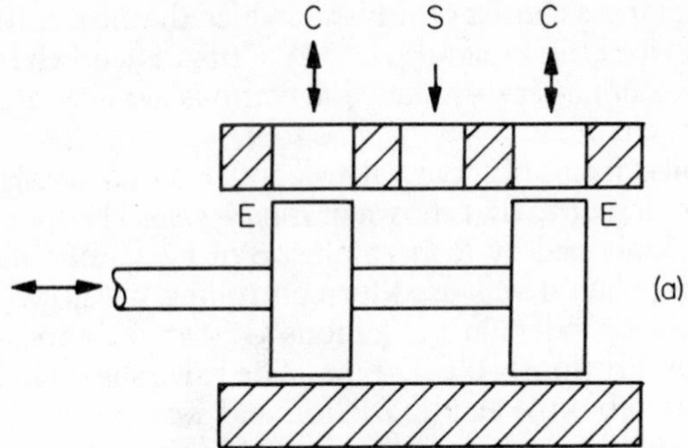

Figure 6.3(a). Two-land spool valve

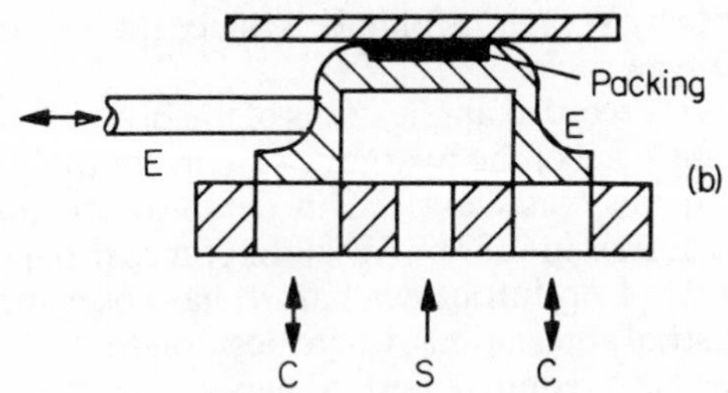

Figure 6.3(b). D-type slide valve

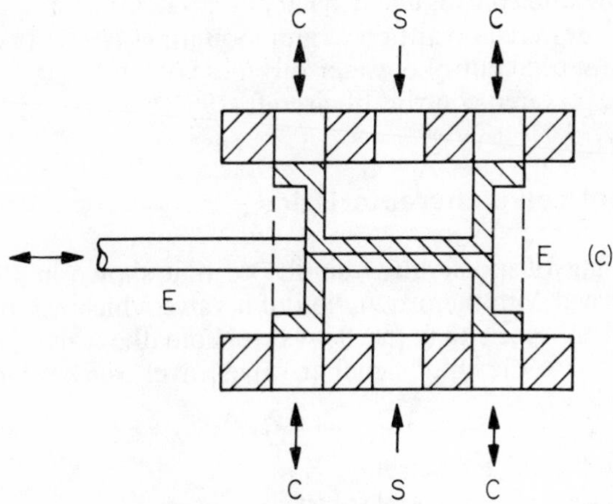

Figure 6.3(c). Back-to-back D-valves
C. Control ports
E. Exhaust region
S. Supply ports

and many of these consist of valves in which the electrically operated opening device is an integral part of the unit. Such valves are called 'electro-hydraulic' servo-valves and various designs are described later in this chapter.

The earliest forms of servo valves consisted of a simple, two-land, spool valve similar to that shown in *Figure 6.3a.* The spool valve was originally developed by Robert Wilson of Patricroft[1] more than a century ago when it was used for controlling the flow of steam to steam engine cylinders at the famous Gorton locomotive works in Manchester. Previously the 'D-type' slide valve shown in *Figure 6.3b* had been used but the erratic friction and wear of the packing had led initially to using two valves mounted back-to-back (*Figure 6.3c*) and eventually to Wilson's idea of forming a solid of revolution by rotating the 'D-type' valve about a tangent to the bow of the D. An earlier paper by Wymer[2] is of interest since it traces the development of the D-type valve.

One of the first recorded applications of the piston type valve was its use in the first stage of the hydraulic system controlling the Eiffel Tower lifts and this was described in detail at the Institution of Mechanical Engineers in 1889 by Ansaloni[3]. Since then many similar systems, often extremely intricate in nature, have been used in a wide variety of industrial applications, where position control, in one form or another, has been required. Several papers have been written on this subject[4] and some of the fields of applications are

(a) Flow rate control in chemical plants (1936)[5].
(b) The powered operation of gun mountings (1947)[6].
(c) The speed control of water turbines (1951)[7].
(d) The powered controls of aircraft (1953)[8].

6.3 Spool valve characteristics

Most designs of spool-valves, similar to that shown in *Figure 6.4*, are concerned with the production of a valve which exhibits linear characteristics such that the flow rate from the valve is directly proportional to the input signal or valve travel. Since we may write

$$Q = f(z, P_v, \gamma)$$

where Q is the valve flow rate,
z is the valve or spool travel,
P_v is the pressure drop across the control ports, and
γ is a parameter relating the area of the port to the valve travel.

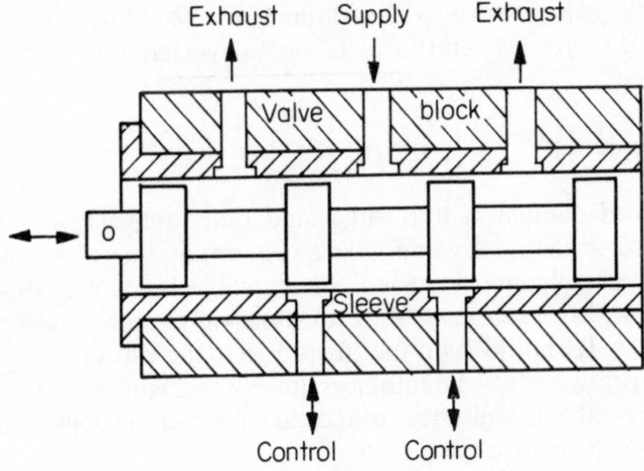

Figure 6.4. *Spool valve*

it can be seen that, if linearity is required, then it is necessary for γ to remain constant which implies rectangular (or fully annular) ports in the valve sleeve rather than circular ones. Circular ports are naturally easier to manufacture, making them popular from a production point of view, but several techniques now exist which facilitate the manufacture of valves having square ports[9]. In practice the economic factor has been sufficiently powerful to ensure that the numbers of designs having circular and rectangular ports are approximately equal and this will probably remain so unless a new method of production for either is devised.

6.3.1 Geometry of ports and lands

For comparison purposes it is useful to define a Reynolds Number for the valve and Khoklov[11], in a comprehensive treatment of this subject, has suggested that this should be

$$R_e = 2\rho V\delta/\mu$$

where ρ is the fluid density;

- V is the mean velocity of the fluid passing through the port;
- δ is the effective valve opening allowing for any radial clearance; (i.e. $\delta = \sqrt{(z^2 + c_1^2)}$ where c_1 is the radial clearance), and
- μ is the fluid viscosity.

Suitable values of the valve dimensions are shown in the table beneath *Figure 6.5* and are those suggested by Conway and Collinson[12].

6.3.2 Valve lap

To avoid confusion it is suggested that 'underlap' or negative lap is used to define a valve where it is impossible to blank off the ports completely, and 'overlap' where valve lands are longer than the port width, such that a 'dead zone' or region of virtually 'no flow' exists. It should be remembered that the physical dimensions of the lap are rather meaningless unless the values of the normal maximum valve travel, the radial clearance and the degree of system filtration are also stated.

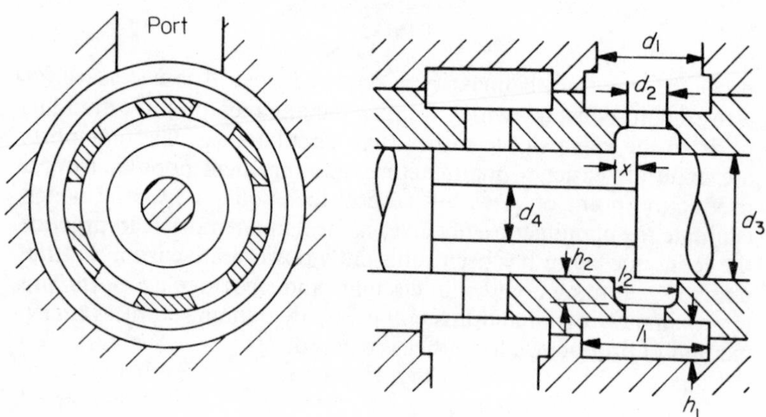

Figure 6.5. Valve dimensions

Assume all sections are to be not less than throat of inlet d_1.
For flow away from throat:

The dimensions h_1 and l_1 must lie between two critical cases, which give $h_1 = 0.4d_1$ with $l_1 = d_1$ and $h_1 = 0.25d_1$ with $l_1 = 1.6d_1$. Suitable dimensions are $h_1 = 0.35d_1$, $l_1 = 1.2d_1$.

Assuming six feed holes, $d_2 = 0.4d_1$.
Dimensions h_2 and l_2 are related as before to d_2, but to feed the valve opening properly h_2 must be greater than x.
Valve opening x is a maximum when $x = 0.25d_3 = 0.25d_1$ approximately.
For flow along annulus of valve spool $d_1^2 = d_3^2 - d_4^2$, that is d_3 is rather greater than d_1.

d_1	d_2	d_3	d_4	x	h_2	l_2	h_1	l_1
10	4	11	5	2.5	2.5	4	3.5	12
10	4	15	11	1.6	2	4	3.5	12
10	4	20	16	1.2	1.5	4.5	3.5	12
10	4	30	26	0.8	1.0	6.5	3.5	12

Although the practical value of valve lap has been the subject of many long and detailed arguments, the question of whether or not it is beneficial and, if so, whether it should be positive or negative cannot be considered settled or to have a single answer. So much depends on the characteristics of the rest of the system and the nature of the required system response that an extensive study is necessary to ensure that all the relevant details have been considered. Briefly, however, it may be said that a very small amount of underlap ensures that the system 'stays alive' provided silting up is avoided and to some extent helps to avoid any tendency of the spool to lock against the liner during long periods of inactivity. At the same time,

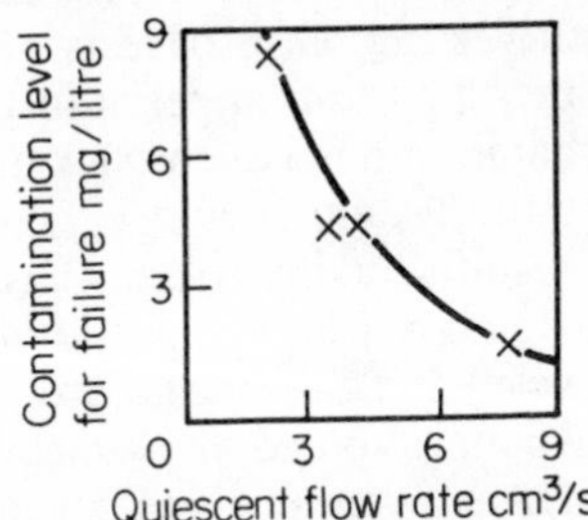

Figure 6.6. Effect of quiescent flow rate on the contamination level producing failure

however, it does entail a quiescent flow which produces a continuous drain on the supply and this is sometimes prohibitive if the latter is limited as in the case of a 'once only' pressurised system, such as that in certain missile applications. Another attraction of underlap is that it tends to increase the stability of systems as is described in section 7.6.2 of Chapter 7 and by Harpur[8], but it is emphasised that it cannot be used to stabilise an otherwise unstable unit. Harpur suggests that a valve having an underlap less than 0.08 mm is liable to silting, but in many systems this amount is equivalent to between 10–20% of the maximum valve travel and the accompanying leakage or quiescent flow would be prohibitive.

A more recent figure for the permissible leakage flow, given by Eynon[9], suggests that the combined underlap and clearance flow should be equivalent to an opening of 0.025 mm for a valve having a maximum travel of 0.8 mm and a radial clearance of approximately 0.025 mm. It is assumed that filtration is achieved with a 4 micron filter which is equivalent to 0.000,160 in. The importance of the effect of contamination level on the desirable quiescent flow rate is illustrated in *Figure 6.6*, which was obtained at the Wright Air Development Centre[13]. It should be noted that these results are not intended to indicate that for large contamination levels the quiescent flow should be kept as low as possible since failure generally occurs when a certain quantity of contamination has built up as is shown by the hyperbolic form of the curve. Details of the effects of asymmetrical lap on the response of a system are given in Reference 14.

When overlap is used the advantages of a virtually zero quiescent flow rate and a 'hydraulically locked' output are obtained. The latter is particularly useful if the output is subjected to any type of random load such as that on an aircraft control surface since to move the output shaft it is necessary to compress the volume of oil contained between the valve control port and the motor or ram piston. Such systems can be made to have extremely high output impedances and Houlobek[15] has analysed and described an interesting series of tests on such a system. The penalty paid with such a system appears in the form of a 'dead zone' which means that in some cases the jack or output does not respond to small inputs to the systems.

Additionally, systems having valves with overlap can be prone to a small but finite, continuous oscillation (generally a form of 'limit cycle') and this occurs even when there is no input. This oscillation can be avoided to some extent by supplying a small but high frequency input signal such that the valve is continually oscillating about its central position, but at a sufficiently high frequency such that it is impossible for the output to follow it. The small input signal and valve movement are known as 'dither'.

6.3.3 General flow rate characteristics

Two flow rate curves are usually considered when assessing the performance of a valve; the most important one being the flow rate/spool displacement and the other being the flow rate/valve pressure drop. A typical example of flow rate/spool displacement is shown in *Figure 6.7a* for a valve having rectangular ports.

For zero valve travel there will generally be a small leakage flow which is mainly a function of the spool and liner clearance as is described below. As the valve travel increases the port area and flow rate generally increase in a linear fashion with it except, of course, in the case of valves having circular ports. For the linear case this relationship can be written as

$$Q = C_d zw\sqrt{(P_v/\rho)} \tag{6.2}$$

where C_d is the discharge coefficient and w is the width of the ports.

It should be noted that the pressure drop across each control port is $P_v/2$ and it is assumed, and indeed confirmed by most experimental

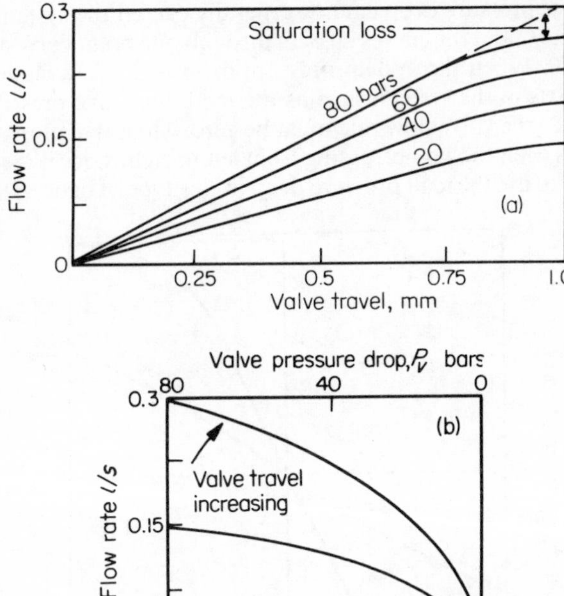

Figure 6.7. (top *Variation of flow rate with valve travel and valve pressure drop;* (bottom) *Variation of flow rate with valve and land pressure drop*

measurements, that the pressure drops across each control port are approximately equal for the same flow rate.

Beyond a certain value of valve travel, the flow rate ceases to increase in a linear manner due to significant pressure losses in other parts of the valve and this is known as flow saturation and is illustrated in *Figure 6.7*.

The slope of the curve, $\partial Q/\partial z$ plays an important part in the response of the system and it is convenient to define it as the valve gain C_z. Hence from equation 6.2 we may write

$$C_z = \left.\frac{\partial Q}{\partial z}\right|_{P_v \text{const}} = C_d w \sqrt{(P_v/\rho)} \tag{6.3}$$

showing that its value is independent of valve travel but a function of the pressure drop across the valve.

The curves in *Figure 6.7b* show the variation of the flow rate with the pressure drop across the valve for various values of valve travel.

The valve pressure drop cannot generally exceed the system supply pressure, P_s, and the curves all pass through one point corresponding to $P_v = 0$. When the system pressure drop is entirely dissipated at other parts of the system, such as the load, the valve pressure drop and hence the valve flow rate must be zero. The valve pressure drop scale has been shown decreasing from left to right since it is common practice to use the load pressure drop in this type of figure. It is again

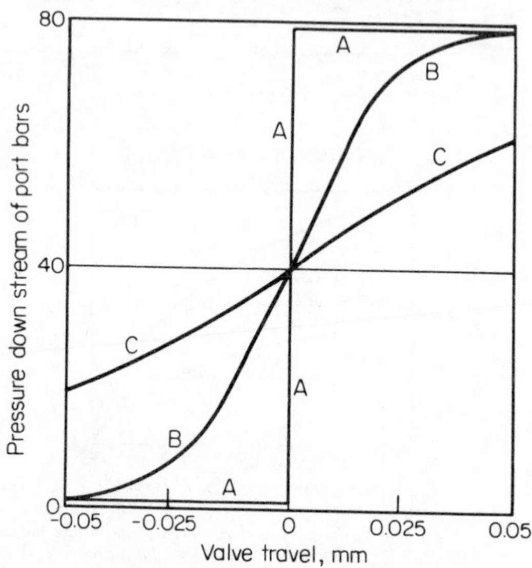

Figure 6.8. *Null pressure gain characteristic* (*supply pressure, 80 bar*)
A. Ideal valve
B. Usual result
C. Excessive lap, clearance or wear.

useful to define the slope of these curves, $\partial Q/\partial P_v$ and this may be written as:

$$C_p = \left.\frac{\partial Q}{\partial P_v}\right|_{z\,\text{const}} = \frac{C_d zw}{2\sqrt{(P_v \rho)}} \tag{6.4}$$

Due to the shape of the curves which is parabolic this derivative, unlike C_z, varies continuously along each curve and is a function of both the valve travel and pressure drop and approaches zero with the valve travel, z.

Another characteristic which is sometimes studied is the 'null pressure gain' characteristic which is obtained when a control port is blocked off and the pressure just downstream of the port but upstream of the blockage is recorded; an example is shown in *Figure 6.8*.

The rate at which the pressure downstream of the port rises with spool travel is a measure of the system stiffness and this a function of the amount of lap and clearance. Ideally, for a valve with both zero lap and clearance, the curve should consist of a vertical and two horizontal lines as shown by lines A in the figure but both lap and clearance effects modify this giving a result of which the curve labelled B is typical. Excessive lap (either positive or negative), clearance or wear would be indicated by the curve labelled C.

6.3.4 Extreme valve openings

The radial clearance between the liner and spool plays a large part in determining the magnitude of the flow rate at small valve openings and Khoklov[11] has obtained the curves shown in *Figure 6.9b* where the vertical ordinate corresponds to the square of the reciprocal of the discharge coefficient. Similar work is also described in Reference 10 and, in addition, Shearer *et al.*[16], MacLellan *et al.*[17] and Viersma[18] have given results of work which has been done on this topic in the USA, England and Holland.

Flow patterns, obtained from a large, low pressure, water model[10] are shown in *Figure 6.9a* and it is convenient to describe the flow through the control port upstream of the load as *outflow* and that at the downstream control port as *inflow*. With outflow conditions and valve opening to radial clearance ratios, (b/d), less than approximately 3.0 the flow continues in a direction parallel to the axis of the spool as is shown by top left hand diagram of *Figure 6.9a*. For larger openings and up to values of the ratio b^2/ac of 0.11 (where the symbols are as shown in the middle left hand figure) the flow leaves the control port at an angle of approximately 69° (see section 6.4.1) but, after passing a separation bubble, is parallel to the vertical wall of the exit chamber. As the valve opening, b, continues to increase beyond the point at which $b^2/ac = 0.11$ the flow breaks away from the vertical wall and remains inclined at an angle of approximately 69° to the spool axis.

The ratio b^2/ac can be considered as that of the opening ratio b/a to the expansion ratio c/b and further details of the characteristics of the separation bubble and its influence on the flow can be found in Reference 17 and methods of detecting incipient cavitation are discussed by McCloy and Beck[19].

A similar set of flow patterns for inflow conditions are shown in the right hand set of diagrams of *Figure 6.9a*. The opening/clearance ratio again plays an important part in determining the type of flow patterns for small valve openings but the transition from attached

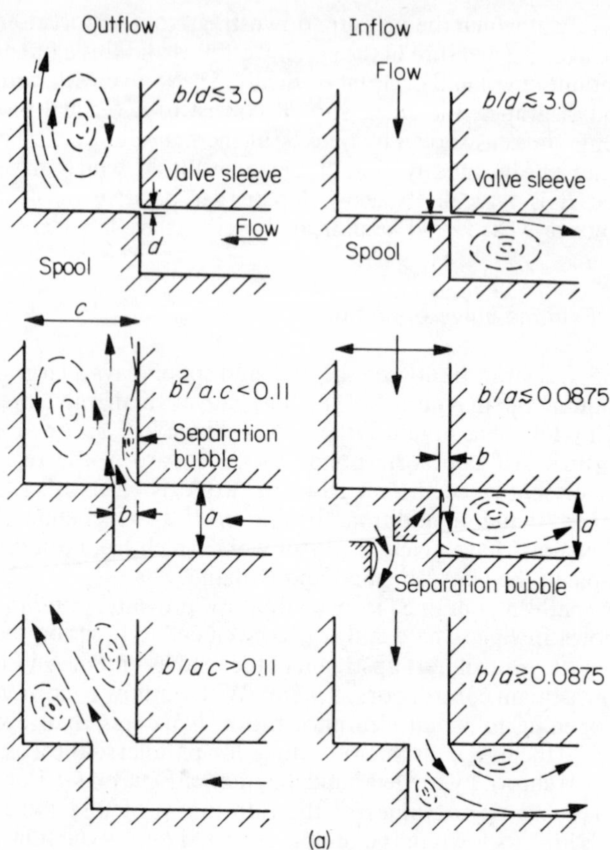

Figure 6.9(a). Variation of flow patterns for outflow and inflow conditions

flow to free jet flow was found[10] to take place when $b/a \approx 0.1$ when the valve opening was increasing. However, the change back from free jet flow to re-attached flow when the valve was closing (i.e. b decreasing) was found to occur when $b/a \approx 0.075$ so that in general terms the change could be said to occur when $b/a \approx 0.0875 \pm 0.0125$ thereby illustrating the presence of a hysteresis region and the independence of the dimension c. Further work on flow hysteresis has been done by McCloy and Beck[20].

In practice, valve designers generally do their best to reduce the magnitude of this region of the valves characteristic by either ensuring that the valve has a large travel such that the small valve openings at which these phenomena occur play only a very small part in valves general operation or, in some cases, by using underlap.

This is not particularly difficult since the opening corresponding to a Reynolds Number of 250 in *Figure 6.9b* is generally no greater than 0.08 mm.

The limit on the maximum permissible valve opening is determined by the losses in the valve passageways and along the valve stem. Some indication of these is provided by Reference 10 but a

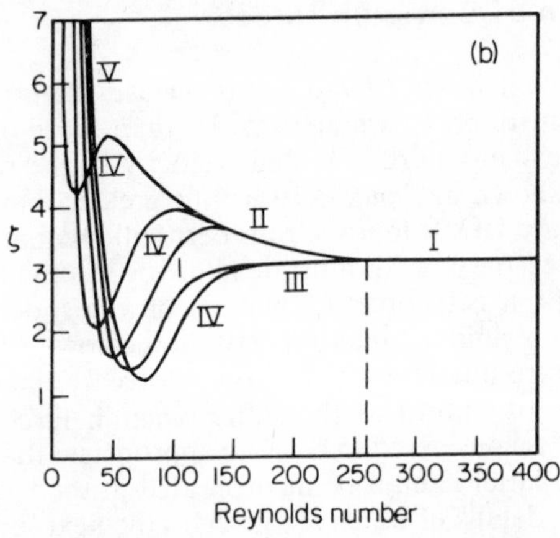

Figure 6.9(b). Variation of discharge at small valve openings

considerable amount of useful research using actual valves remains to be done in this field. Naturally, the maximum area of the control ports must be smaller than the cross-sectional area presented to the flow at all other points in the hydraulic circuit and in the absence of any detailed information a factor of at least five to ten times smaller would seem to be desirable.

6.3.5 Linearisation of flow/travel characteristic

If it is assumed that the valve travel will be proportional to the input signal then the use of rectangular ports is to be preferred if linearity between flow and input signal is desired. Alternatively, a series of small staggered circular ports can be used and indeed has been recommended by Clark[21] in an attempt to reduce other non-linear effects as described in section 8.4.3 and this method is compared with others by Carrington[22].

A novel method of linearising the overall flow rate/input signal characteristic so that it is independent of the loads pressure drop by using the valve reaction force has been suggested in Reference 10 and the method has been examined in more detail by McCloy and Martin[23]. It consists of utilising the reaction forces in conjunction with a spring attached to the spool so that the flow rate can be made virtually constant over a wide range of the valve pressure drop and this is described in detail in the next chapter, section 7.6.8.

6.4 Reaction of Bernoulli forces

When a spool valve (see *Figure 6.4*) is opened then, provided the area of the control ports is much smaller than those of the supply port and the exhaust ports, the fluid velocity in the region of the land face AB shown in *Figure 6.10* will be greater than that in the region of the face CD. It follows from Bernoulli's equation that the static pressure on the face AB must therefore be less than that on CD and, as a result, a nett force, F_o, will act on the spool tending to close the valve. A similar condition exists at the downstream control port and again results in a 'valve closing' force F_i and the total of these forces, F, is known as the static reaction force. (The term 'reaction force' is considered to be more appropriate than 'Bernoulli force' since the latter cannot be incorporated in the term 'dynamic reaction force', details of which are given in the next section.)

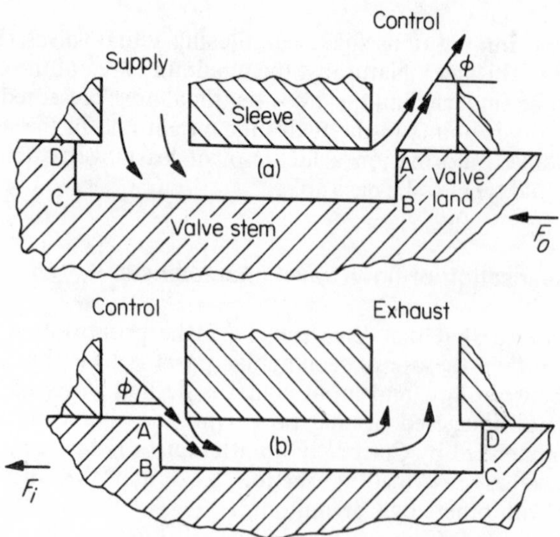

Figure 6.10. Valve reaction forces. (a) Outflow port, supply to control; (b) Inflow port, control to exhaust

Most methods of reducing the reaction force depend on methods of changing the sign of the inflow portion of the force such that it cancels out a significant part of the outflow portion of the force.

6.4.1 Static reaction force equations

The first work specifically on the reaction force of spool valves was probably that of Swain[24] in 1942 and this was followed by the work of Sweeney[25] on an Admiralty valve. It remained for Lee and Blackburn[26] in 1951, however, to devise the following method of calculating the value of the force.

Since the area of the control port is generally much less than that of the supply port the velocity and hence momentum of the fluid is much greater at the former. In addition, it has been shown[10] that the flow is virtually radial at the supply or exhaust port whereas at the control port it is inclined at a mean angle, ϕ, of approximately 69°, to the spool axis. As a result, the rate of change of momentum of the fluid in the axial direction is given by:

$$(\text{mass flow rate}) \times (\text{velocity}) \times (\cos\phi)$$

or

$$F_o = \rho Q \frac{Q}{AC_c} \cos\phi \tag{6.5a}$$

where A is the control port area and C_c is the contraction coefficient.

In a classical, theoretical study of the flow of jets from apertures, von Mises[27] has shown that the discharge angle for configurations similar to those of the control ports is 69° for the limiting case of zero opening and that it does not change greatly for openings similar to those used in practice. In addition he showed that the theoretical value of the contraction coefficient at small openings is approximately 0.67 and his results are summarised in *Figure 6.11*. Since the velocity coefficient may generally be taken as nearly unity and the discharge coefficient is then given by $C_d \approx C_c \approx 0.67$, the total reaction force, F, is given by

$$F = (F_o + F_i) \approx \frac{2\rho Q^2 \cos\phi}{AC_d} \tag{6.5b}$$

Since

$$Q = AC_d \sqrt{(P_v/\rho)}$$

and for a single port $A = wz$, this then gives

$$F \approx 2P_v wzC_d \cos\phi$$

or

$$F \approx KP_v z \tag{6.5c}$$

where $K = 2wC_d \cos\phi$.

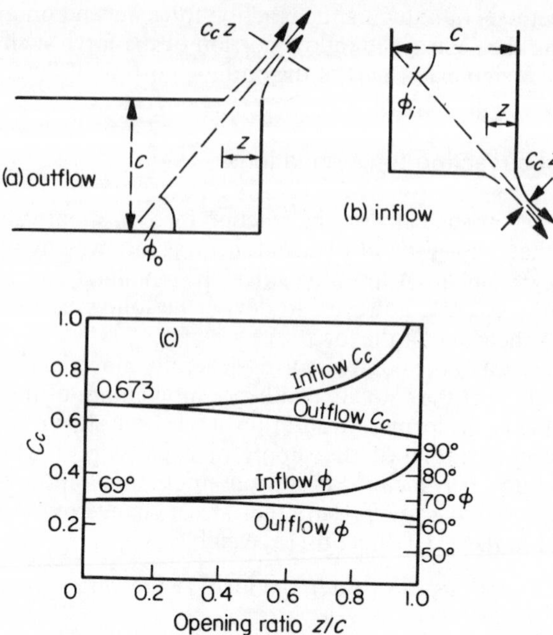

Figure 6.11. Theoretical contraction coefficients and discharge angles as derived by von Mises

It should be noted that for a constant pressure drop across the valve the reaction force increases linearly with the valve travel, z, but if the valve pressure drop varies during this travel then allowance must also be made for this effect as is discussed later in section on valve instability. The force, given by equation 6.5, is the 'static' reaction force since it acts at all times on the stationary valve spool if the valve is open. Even when air or gas is used as the working fluid, the static reaction force can be significant, and some interesting results of its measurement have been obtained by Feng.[28]

6.4.2 The dynamic reaction force

The term dynamic reaction force is used to describe an additional force also examined by Lee and Blackburn which is associated with the acceleration of the fluid surrounding the spool stem as shown in *Figure 6.12*. If the valve travel is suddenly increased then the flow rate will also increase and the mean axial velocity of the liquid around the spool will rise. As a result there must be a reaction on this fluid together with an equal and opposite one on the land faces and for the valve shown the total force, F_D, will be given by

$$F_D = \rho a' L_1 \frac{du_1}{dt} - \rho a' L_2 \frac{du_2}{dt}$$

where a' is the area of the annulus formed by the valve stem and liner and u is the mean velocity across this area.

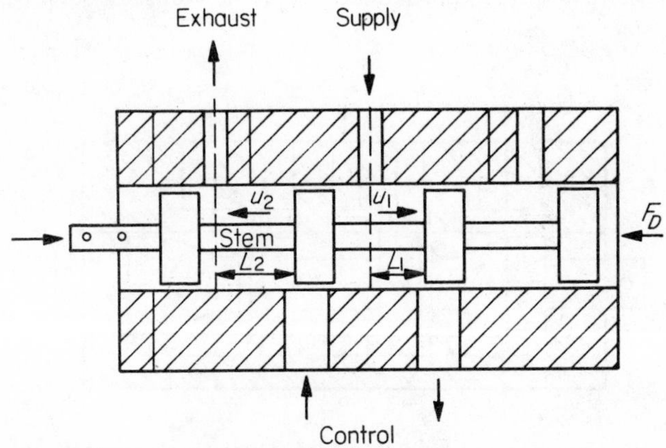

Figure 6.12. The dynamic reaction force. The valve shown would tend to be unstable since $L_2 > L_1$.

Since the velocities u_1 and u_2 in the two portions of the valve will be equal but in opposite directions the nett force may be written in terms of the flow rate Q ($Q = u_1 a' = u_2 a'$) giving

$$F_D = \rho \frac{\delta Q}{\delta t} \{L_1 - L_2\} \tag{6.6}$$

showing that a force tending to increase the valve opening will act on the spool as the valve opening is increased if $L_1 < L_2$. Such a situation is most undesirable and several early designs of valve were unstable for this reason.

At one time a simple 'cure' consisted of reversing the roles of the supply and exhaust ports but this is undesirable since there are then two high pressure regions and there is only one land between the supply pressure and the atmosphere. Now that the dynamic reaction force is recognised it can always be made to provide positive rather than negative damping by a suitable choice of the lengths L_1 and L_2. The term $(L_1 - L_2)$ is generally referred to as the 'Damping Length' since it provides a damping force (i.e. one which is proportional to velocity) on the spool.

6.4.3 Reaction force compensation

In addition to describing the mechanics of both the reaction forces in detail, Lee and Blackburn also suggested a valve spool and liner geometry that would have virtually no nett static reaction force. This is naturally very desirable because for high pressure systems and large fluid powers the force required to hold the valve open can be considerable and often amounts to many newtons. In such cases it

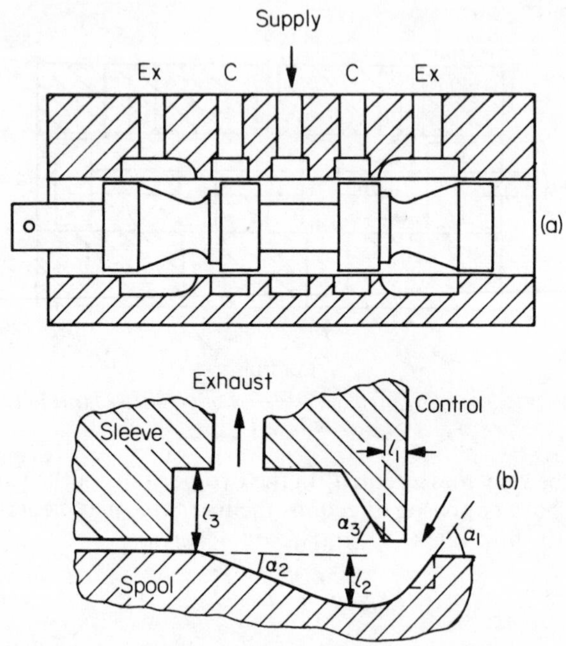

Figure 6.13. Lee and Blackburn force compensated valve. (a) Spool and sleeve; (b) Detail of spool and sleeve between control and exhaust ports.

is necessary to use large valve operating devices which are both costly and heavy. The latter factor is naturally important in applications where minimum weight is essential and it is often better to have a valve operated by a simple electro-mechanical device driven in turn by a small amplifier rather than a valve requiring a large solenoid which in turn would require a large operating current and which would have a large time constant.

The configuration suggested by Lee and Blackburn of the MIT is shown in *Figure 6.13* and with the optimum values of the various parameters, it gives excellent results since the sign of the inflow component of the static reaction force is changed such that the total static reaction force, F, is virtually zero. Whilst the improvement is excellent, the cost of producing the valve is approximately doubled and because the price of even a simple four landed valve is generally between £50 and £100 the method, which involves removal of part of the inner surface of the liner, is not always acceptable.

The decade following the publication of their work saw intense activity in several laboratories where workers were attempting to

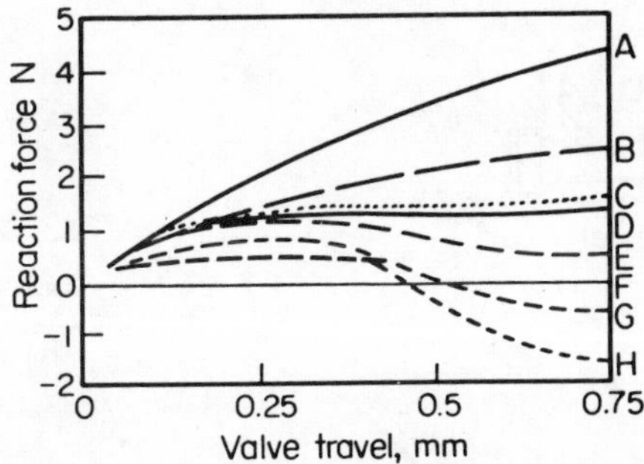

Figure 6.14. Effect of valve dimensions on reaction force

discover alternative and cheaper methods of reaction force reduction and the work of Eynon is worthy of particular note. His results[9], published several years after they were obtained, give valuable design data concerning the various parameters used by Lees and Blackburn as illustrated in *Figure 6.14* and described in *Table 6.1*.

One of the production difficulties concerning the MIT valve was the blending of the curved profile of the stem to the land and this is made much simpler if a small step at the junction is permissible as is shown in *Figure 6.13*(*b*) and by the broken lines in *Figure 6.14*.

Table 6.1 DETAILS OF VALVES USED TO OBTAIN RESULTS SHOWN IN FIGURE 6.14
(Valve diameter 6.35 mm. Dimensions as shown in *Figure 6.13b*)

Curve	*Recess Depth* l_3 (mm)	*Spool Recess* l_2 (mm)	*Recess Chamber* α_3°	*Valve Taper* α_2°	*Liner Land* l_1 (mm)
A	2.1	1.16	60	10	0.91
B	2.38	1.16	60	20	0.76
C	2.44	1.16	60	20	0.64
D	2.44	1.42	45	15	0.64
E	2.44	1.16	45	20	0.64
F	2.44	1.29	45	15	0.64
G	2.44	1.16	45	20	0.64
H	2.44	1.16	45	15	0.64

It was found that the degree of compensation was very dependent on the size of this step since it naturally had a great influence on the flow pattern. Eynon also considered other methods of force reduction and in particular showed that a degree of reaction force compensation could be achieved by simply tapering the two outer portions of the stem of the valve spool as shown in *Figure 6.15*.

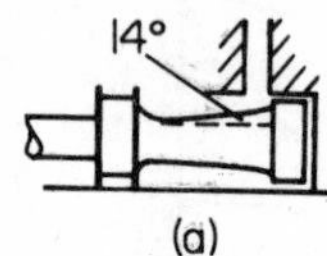

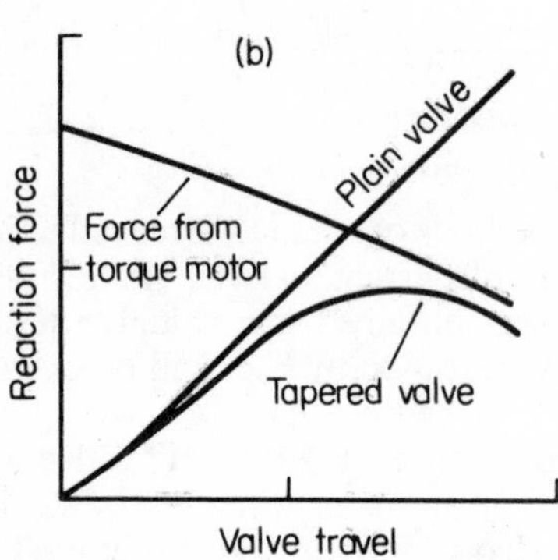

Figure 6.15. Force compensated valve. (a) Configurations; (b) Force/valve travel

Parallel work at Cambridge[17,29] produced a simple design generally known as the 'thick stem valve' and which, although costing no more to produce than a standard spool valve, achieved approximately 80% reduction of the reaction force. It is shown in *Figure 6.16* where it can be seen that if the technique of increasing the diameter of the outer portions of the stem is taken too far, then the flow is throttled along the thickened portion of the stem and the required maximum flow rate cannot be obtained. Work by Clarke[21] also produced a similar design and, in addition, an alternative method in which the control ports were replaced by a series of small staggered holes. This latter design is shown in *Figure 6.17*, but again, the accuracy required to position the holes

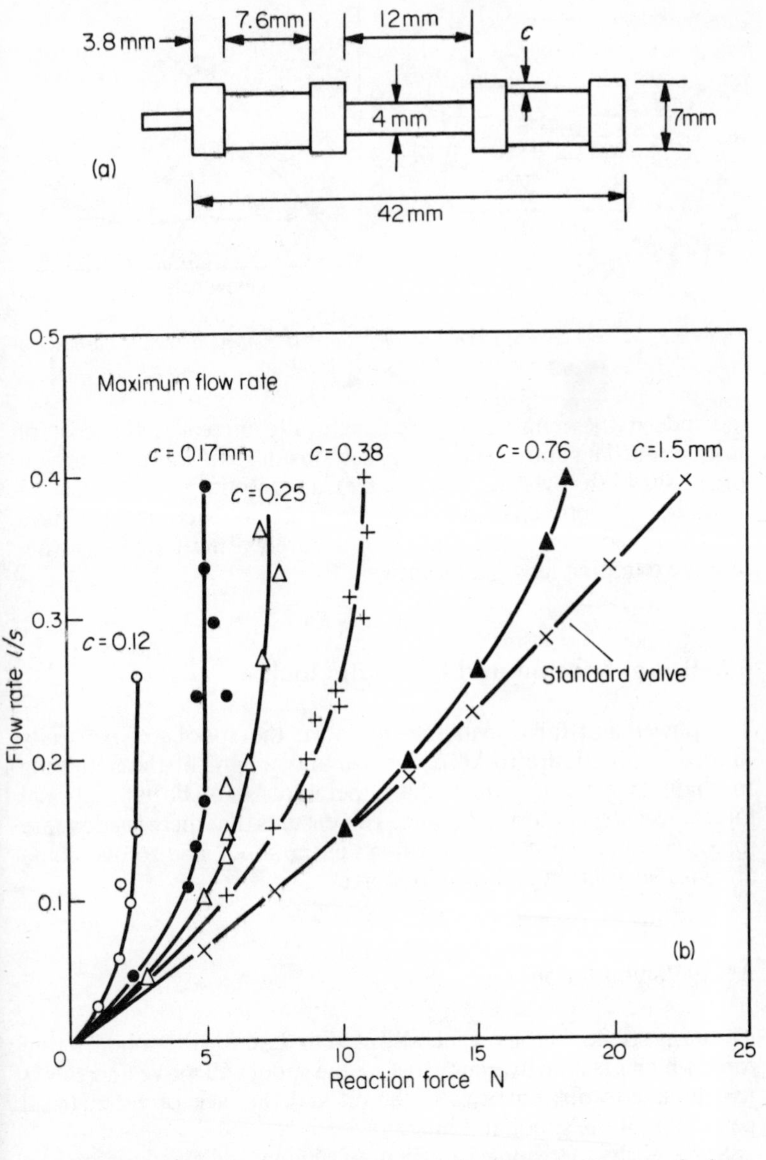

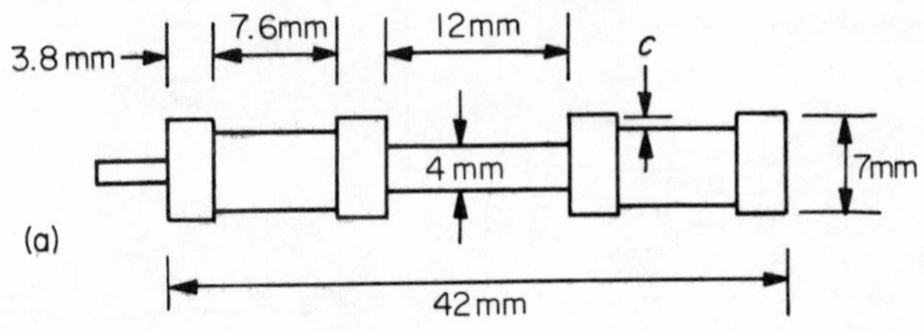

Figure 6.16. Reaction force compensation by means of thick-stemmed valve
(a) Diagram of thick-stemmed valve spool
(b) Reaction force/flow relationship. Pressure drop 100 bars

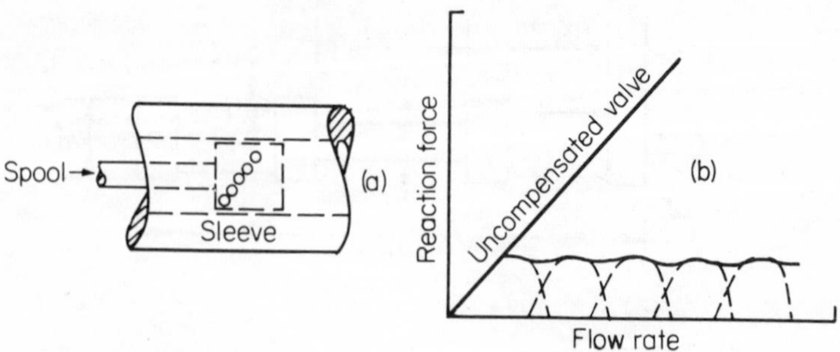

Figure 6.17. Valve with 'staggered porting'. (a) Port configuration; (b) Variation of reaction force with flow rate

and indeed the number required, naturally increases the cost of production. The above and other methods of reducing the reaction force have been reviewed in detail by Carrington[22].

Studies of methods of reducing the dynamic reaction force have not been made since its elimination is purely a matter of ensuring that the damping length is negative.

6.5 Spool friction and hydraulic lock

The power and force available to move the spool are generally limited by the desire to keep the size and weight of the actuating mechanism to a minimum, and it is therefore essential that frictional forces are also kept to a minimum. This ensures that there is adequate force available to accelerate the mass of the spool, and to overcome any viscous friction and reaction forces.

6.5.1 Valve friction

The term 'stiction' is now generally used to describe the solid friction force which must be overcome before the spool will move and this is usually due to dirt particles in the oil and the lack of geometrical perfection of the spool and liner.

Some early work on the effects of sludge and fine dirt on the performance of a valve spool mechanism is described by Alcock[30] who found that the valve operated satisfactorily even though the

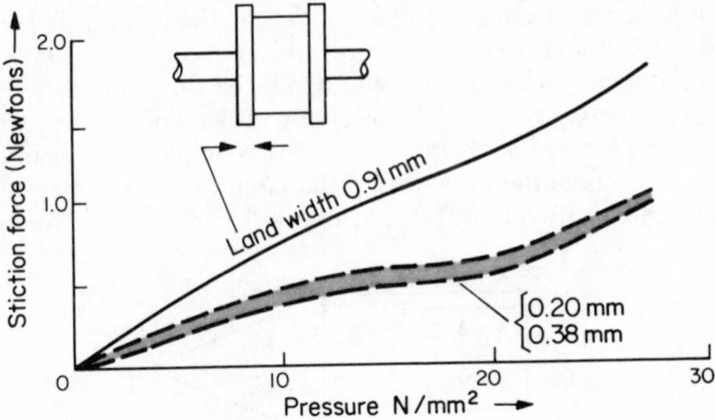

Figure 6.18. Valve stiction curves obtained by Eynon.[9] *Results for land widths of 0.2 mm and 0.38 mm lie within shaded band*

contaminent particle size was several times greater than the radial clearance between the valve spool and sleeve, provided no metallic chips or burrs such as those off screw threads were present. Lack of lubricity of the fluid can also cause poor operation, but another effect generally known as 'hydraulic lock' is of greater importance and is discussed below. A simple method of reducing the stiction effects is to use very narrow lands and a method, suggested by Eynon[9], is shown in *Figure 6.18*, together with the corresponding 'stiction' curves.

6.5.2 Hydraulic lock

Considering an isolated spool land such as that shown in *Figure 6.19a* (and assume for simplification that the flow is two dimensional) it can be seen that the force resulting from the fluid pressure acting the upper and lower surfaces is independent of the eccentricity of the land and liner. As a result, there will be no tendency for land to move in the vertical plane. Indeed if the land tilts then a righting moment will act on it tending to restore it to a position parallel to the liner bore although at the same time the vertical forces will be out of balance.

If, however, the land is tapered such that the 'arrowhead' formed by the continuation of its surface points in the same direction as the flow produced by the pressure gradient (*Figure 6.19b*) then a displacement of the land from its central position will result in pressure profiles which give rise to a net lateral force in the same direction as the displacement. As a result, the land will become forced onto the liner surface and this condition is known as 'hydraulic lock' since the land will be locked onto the liner by the static pressure of the fluid. If the direction of the taper or the pressure gradient is reversed such that the 'arrowhead' points towards the high pressure then the direction of the lateral force is reversed and this condition is therefore stable insofar as the force tends to restore

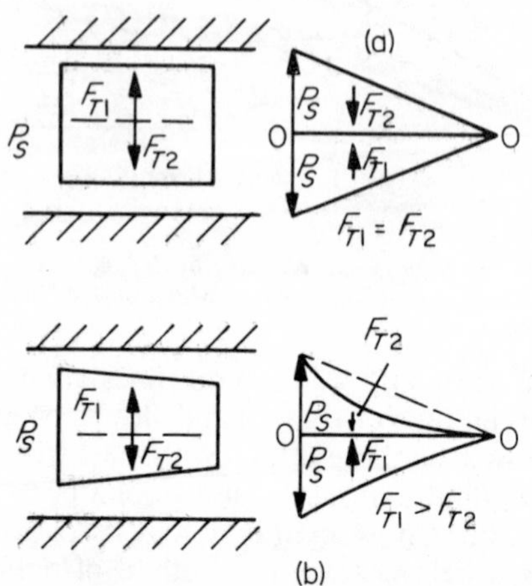

Figure 6.19. (*a*) *Parallel land and pressure distribution;* (*b*) *Tapered land and pressure distribution*

the land to a central position. This latter condition is naturally desirable. The magnitude of the transverse force has been examined experimentally[25] and Royle[31] has calculated its magnitude, F_T, showing that it is given by the equation,

$$F_T = \frac{\pi}{4} LDP_1 G$$

where the notation is as shown in *Figure 6.20*.

When hydraulic lock was first discussed Stringer[32] suggested that it might well be due to the silting up of the fine clearances by minute particles of dirt in the oil but although this is naturally a contributory factor, Alcock's results[30] suggest that the explanation shown in *Figure 6.19* is often the correct one. Excellent experimental confirmation of the nature and magnitude of the hydraulic locking force was provided by a series of tests by Whiteman[33] on a liner and some spools which were deliberately machined to have tapers

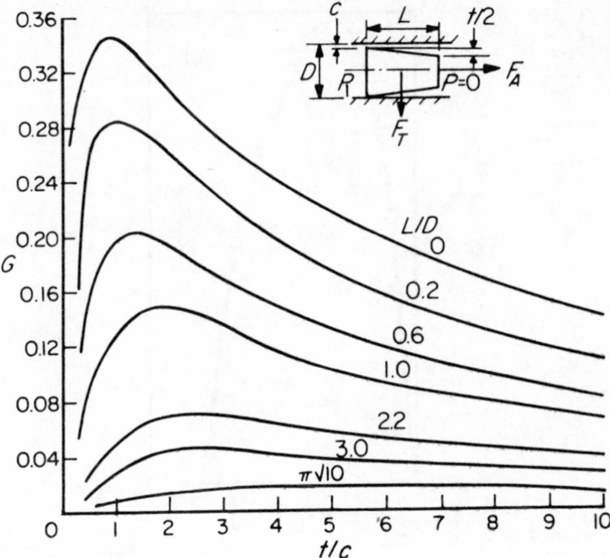

Figure 6.20. Transverse hydraulic locking force on a piston

giving a locking force. A further feature of these tests concerned the condition when the liner is arranged such that it may bellow outwards under fluid pressure. If this situation exists then a pressure will be reached at which the locking force is a maximum and eventually, if the pressure is raised sufficiently beyond this point, the locking force will disappear. The results of one of these tests is shown in *Figure 6.21a*.

Two methods of reducing or eliminating hydraulic lock are now in general use, the first involving the machining of a series of small, circumferential grooves around each land. This prevents an excessive amount of pressure difference occurring around the circumference of the land and in addition provides small traps for any dirt particles in the fluid. Although increasing the number of grooves progressively lowers the locking force, the amount of improvement obtained after the first three or four is generally insignificant and it is generally not advisable to use more than this number because of the accompanying increase in the leakage rate.

The second method consists of deliberately using tapered lands although the amount of taper permitted to eliminate hydraulic lock must generally be kept small again in order to avoid an unacceptable level of leakage. A sometimes attractive alternative to this second method is to machine one or more steps along the lands giving a

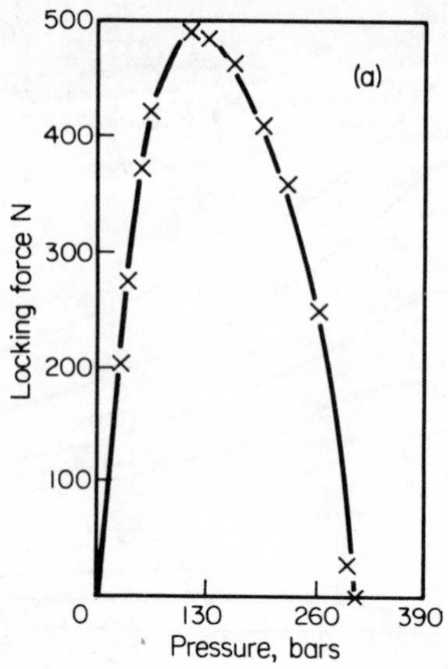

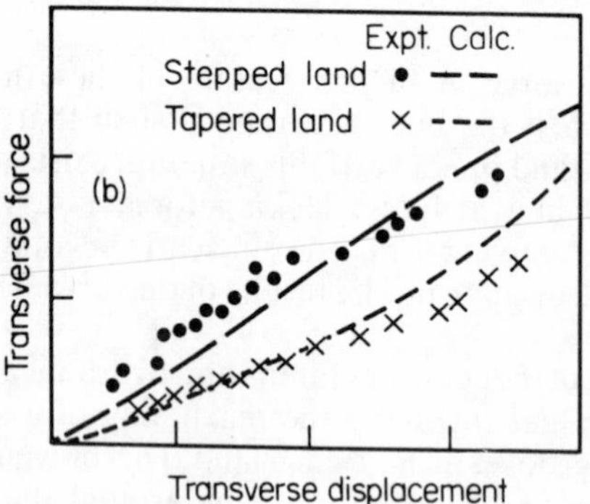

Figure 6.21. Experimental results. (a) Variation of hydraulic locking force with pressure[26]*; (b) Variation of transverse force with transverse displacement*[25]

step-like approximation to the desired non-locking taper, and some interesting results concerning the relative merits of the deliberate tapering/stepped land techniques have been obtained by Weber[34] who measured the centralising lateral forces and the corresponding lateral displacements of two spool valves having these geometries. The results of two of his tests are shown in *Figure 6.21b* and the agreement between the experimental points and the calculated curves is very satisfactory. It was suggested that the discrepancies between the experimental and the theoretical results might be attributed to the fact that the laminar, two-dimensional flow regime is not fully established immediately the fluid enters the small clearance between the spool land and sleeve and that the small peripheral component of the flow past the land was neglected in the derivation of the theoretical curves.

6.6 Valve designs

There are now so many designs of various electro-hydraulic servo-valves being manufactured in both Britain and abroad that it is sometimes difficult for those concerned with such devices to discover what special features distinguish one from another. As a result, a method of classifying the various designs has been devised and is described later in section 6.7.1 of this chapter.

The impetus given to research in the field of automatic control by the 1939–45 war continued long after hostilities ceased and considerable progress was made particularly in the United States of America at both the Massachusetts Institute of Technology (MIT) and at the Cornell Aeronautical Laboratory. At the former a team, which included Drs. Lee and Blackburn of the Dynamic Analysis Control Laboratory (DACL) was responsible for the significant improvements in both the design and analysis of the spool valve and its characteristics as described in earlier sections of the chapter.

Various other single-stage types of valve were designed and it would appear that it was only later that any great effort was directed towards the development of a two-stage valve and this resulted in the use of the mechanical feedback of a flow-rate signal. In parallel with the work at the MIT the development programme at Cornell quickly resulted in the two-stage, valve design of which the first stage is now generally referred to as the 'nozzle-flapper' system. A member of the team, Moog, realizing the potentialities of this device, subsequently formed a valve manufacturing company and can now probably claim to be the world's largest supplier of electro-hydraulic control valves. His designs are manufactured under licence in Britain.

Whilst commercial success has obviously favoured the American designs a certain amount of progress has been made in Britain, much of which was the work of the late J. Ford of the Admiralty. In addition, work at the Royal Aircraft Establishment during the period 1950–60 by Eynon has resulted in the development of a wide range of single-stage spool valves and this was followed by a two-stage nozzle-flapper design. Novel contributions from other countries include the earlier work on the valves for the German V2 missile control system and the fairly recent French design of a two-stage nozzle-flapper valve by the late Dr. C. R. Himmler at the Centre Recherches Hydrauliques et Electriques (CRH) in Paris. The latter design is manufactured by Société d'Optique et de Mécanique (SOM) of Paris and is also in production in Britain. The early design of a spool valve by the Société d'Applications des Machines Motrices (SAMM) also of Paris, is worthy of note and this was used successfully on the 80-tonne flying boat, Laté 631 in 1940 with a supply pressure of 50 bars[35].

6.7 Basic designs

It is possible to classify most types of electro-hydraulic control valve by their basic designs, which depend mainly on their principles of operation and, to a lesser extent, on their geometry[36]. The prime function of most valves is to meter the flow of fluid, supplied at a high pressure, to either a linear (jack or actuator) or rotary (motor) hydraulic machine and Hadekel[37] has provided an interesting account of some of the many systems in which they are used. The speed of the machine or, as it is often called, the load is generally proportional to the flow rate from the valves, which, with the exception of some single-stage designs, usually have a spool valve as their output stage.

6.7.1 Classification of valve designs

The following method of classification has been adopted and in general follows that used in earlier work[36].

(i) Single-stage valves:
- (a) Spool.
- (b) Rotary spool.
- (c) Split spool.
- (d) Sliding plate.
- (e) Rotary plate.
- (f) Askania.

(ii) Two-stage valves:
- (a) Double spool.
- (b) Nozzle-flapper: spool.
- (c) Double nozzle-flapper: spool.
- (d) Plate: spool.
- (e) Askania: spool.

(iii) Three-stage valves: this type of valve is by no means as common as the single-stage and two-stage valves and in general consists of a two-stage design with an additional spool added to provide the output stage. For this reason no classification of this type of valve is given.

6.8 Single-stage valves

6.8.1 Spool-type valve

The simplest form of valve to control both the direction of flow and flow-rate is the three-way valve, but since this will control the flow to only one side of the load a differential area ram must be used. A more common design is the four-way valve, shown in *Figure 6.22a*. The symmetry of this valve results in improved linearity[38]. However, there are three critical axial dimensions of the spool, *a*, *b*, *c*, and also of the sleeve, *d*, *e*, *f*, and the cost of holding these to the necessary limits can often be excessive. To alleviate these manufacturing difficulties designers frequently use alternative designs to the four-way spool some of which are described below.

6.8.2 Rotary spool valve

This type of valve, the earliest designs of which date from Roman times, is not commonly used in servo-systems. In its basic form it is no easier to make than the four-way spool type, but the general configuration shown in *Figure 6.22b*, does permit the use of low friction bearings.

6.8.3 Split-spool valve

To reduce the number of critical axial dimensions of the four-way valve, two separate three-way valves may be linked together as shown in *Figure 6.22c*, and a means of zero adjustment can be provided in the links. Although manufacturing problems associated with porting are reduced, further difficulties are introduced in that two parallel bores have to be made, and there may be additional backlash in the linkage. Furthermore, its weight is usually greater than that of the simple four-way valve, which in some applications is unacceptable.

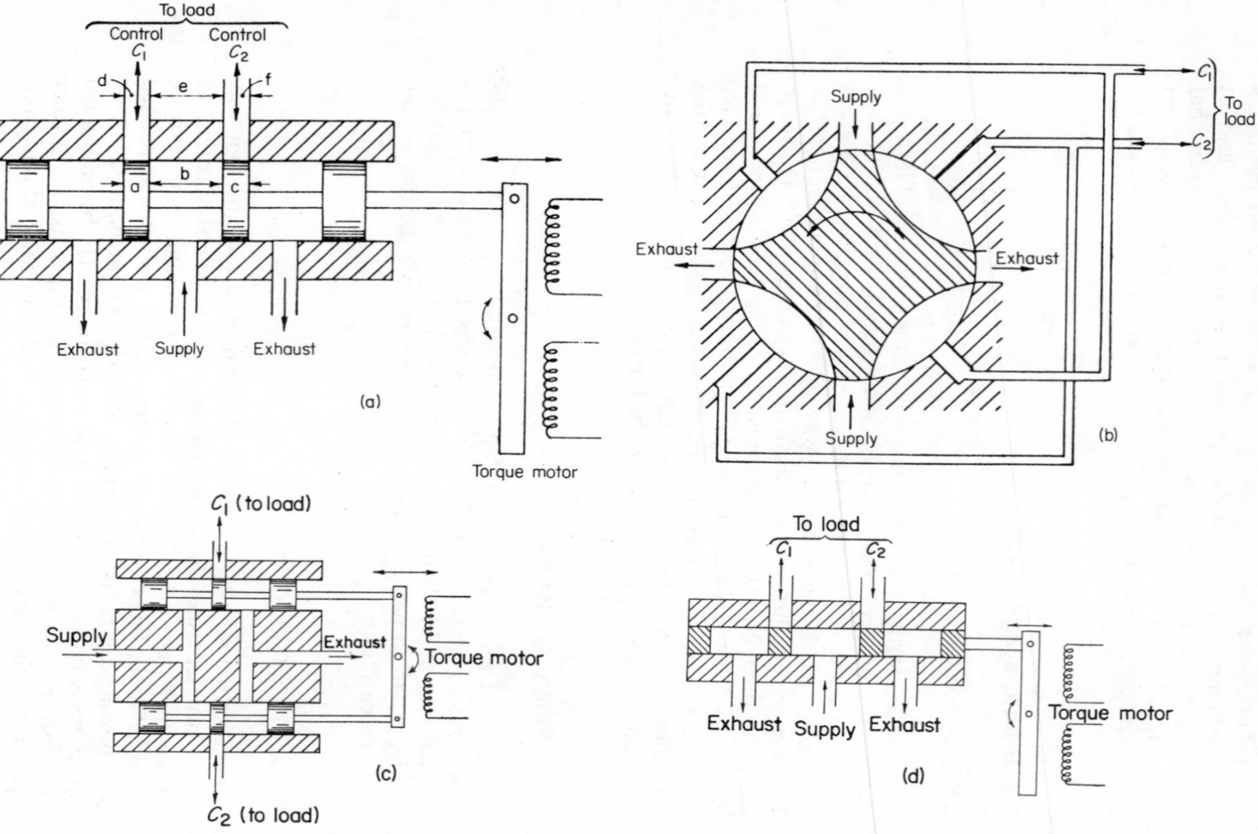

Figure 6.22. (a) *Spool-type valve*; (b) *Rotary spool valve*; (c) *Split spool valve*; (d) *Sliding plate valve*.

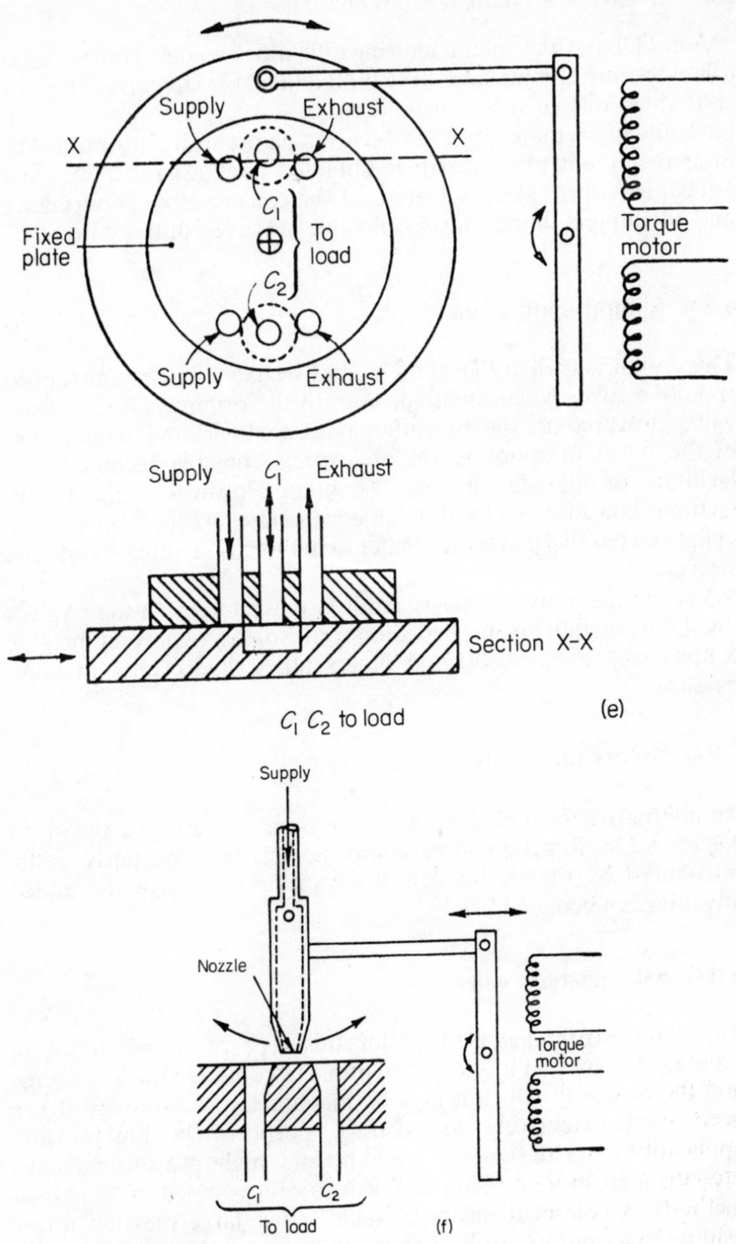

Figure 6.22. (e) Rotary plate valve; (f) Askania-type valve

A design in which manufacturing difficulties associated with axial tolerances are eliminated is the Elliott adjustable lap valve. This is a split, three-way valve in which the two spools are mounted back-to-back and are actuated by a central arm against spring restraint. The desired amount of valve lap is obtained after assembly by axial adjustment of the sleeves. Details of the construction, performance and advantages of the split spool valve are given fully by Besser[39].

6.8.4 Sliding plate valve

This design, shown in *Figure 6.27d*, may be likened to an unwrapped spool (i.e., two-dimensional), or even to the original D-type steam valve. It overcomes the difficulties associated with the manufacture of the bores in spool valves and special porting techniques to facilitate its manufacture may be used. However, some manufacturers consider that the difficulties associated with the production of flat and parallel plates are greater than those of making spools and sleeves.

Various methods are used to reduce friction forces; in some valves the sliding member is suspended on spring plates to prevent metallic contact (see *Figure 6.29*) and others utilise hydrostatic pressure balance.

6.8.5 Rotary plate valve

An alternative form of the plate valve is the rotary type shown in *Figure 6.22e*. Reaction force compensation may be fairly easily introduced by the use of deflector vanes which have the added advantage of being adjustable.

6.8.6 Askania-type valve

The main advantages claimed for this type of valve, shown in *Figure 6.22f*, are that it is less susceptible to contamination clogging, and the ease with which it may be manufactured. Although it has been used extensively for control purposes in low-pressure applications, its design for medium and high pressure systems presents a far more difficult problem since it is based on empirical methods. A comment has been made[38] that large reaction forces leading to serious instability can occur, although the reasons for this are not known. This valve, the action of which depends upon the conversion of kinetic energy of the jet into static pressure at the spool or ram, is referred to in the United States as the 'jet-pipe' design. However, it is probably preferable to avoid this term to prevent confusion with other nozzle or jet designs used in two-stage valves.

6.9 Two-stage valves

Two-stage valves were designed to overcome the practical limitations of power and response of single-stage valves. Any of the previously mentioned single-stage valves may be used as the first stage of a two-stage valve with a spool or a plate as the second stage, and the following are examples.

6.9.1 Double spool valve

In this valve, shown in *Figure 6.23a*, the output from the first stage is used to control the motion of the second-stage spool and hence produces delivery to the load. Feedback from the output stage is

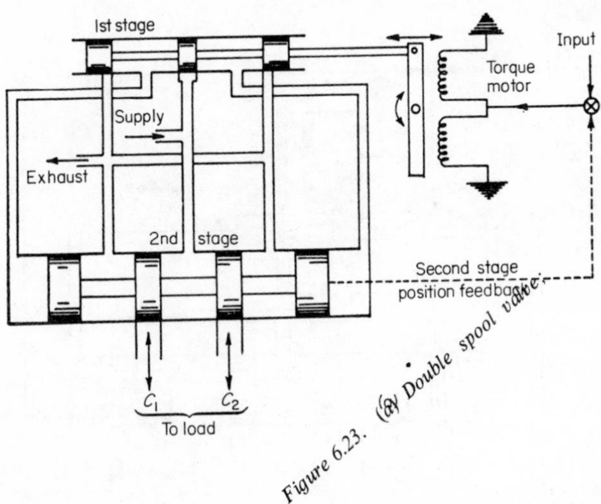

Figure 6.23. (a) Double spool valve.

necessary to avoid the integrating effect of the two spools and is usually accomplished either electrically or mechanically (e.g. *Figure 6.26a*).

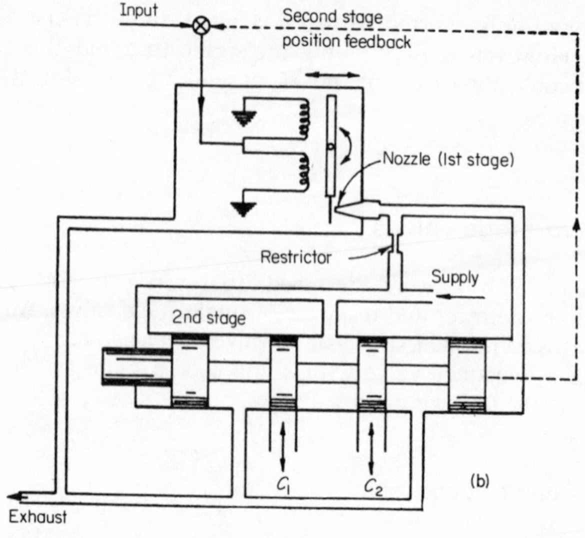

Figure 6.23. (b) Nozzle-flapper spool valve.

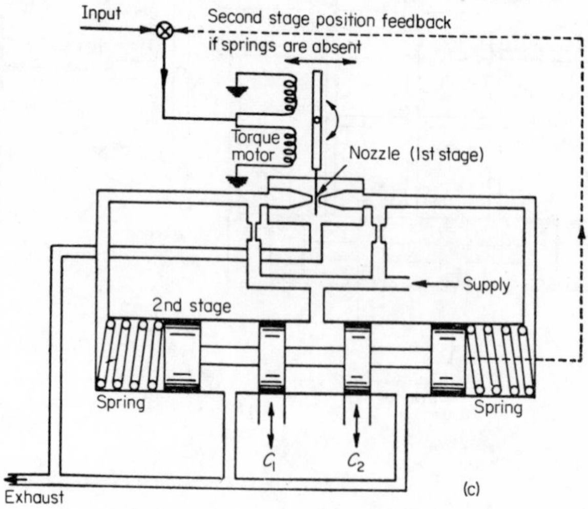

Figure 6.23. (c) Double nozzle-flapper spool valve

6.9.2 Nozzle-flapper spool valve

The basic configuration of the single nozzle-flapper valve is shown in *Figure 6.23b*, where it can be seen that a movement of the flapper towards the nozzle causes an increase in the pressure in the chamber between the nozzle and the restrictor which causes the second-stage spool to move and thus delivers oil to the load. Movement of the flapper in the other direction will reduce the chamber pressure causing the spool to move in the opposite direction and produce reverse motion of the load. Thus the nozzle and flapper act as a variable impedance and a variable but controlled amount of the supply pressure acts on the right-hand end of the second-stage spool. A very extensive study of the characteristics of this type of valve has been made by Kinney *et al.*[13]

6.9.3 Double nozzle-flapper spool valve

Representing a further development of the design previously described, this valve is shown in *Figure 6.23c*. The flapper and nozzles control the pressures at both ends of the second-stage spool and thus the operation of the first stage may be likened to that of the four-way valve. Although a fairly high quiescent flow is inevitable with this design the power loss is usually not significant for most applications.

The introduction of the nozzle-flapper device as the first stage, with its low inertia and short stroke was a major contribution in the field of two-stage valve design. The design shown in *Figure 6.23c*, is basically that of a Moog valve and was one of the earliest of a large series of valves having a nozzle-flapper system as the first stage.

6.10 Other single-stage valves

Although plate valves are no easier to produce than spool valves a fairly simple refinement to the production technique of the former facilitates the manufacture of a zero-lap design. details of the modified machining process are given by Lee[40]. Adjustable depth baffle plates (deflectors) are often used to alter the effect of reaction forces in these valves and a remarkably high degree of performance tuning can be achieved if special geometries are used.

An example of a swinging-plate valve is shown in *Figure 6.24* and this design is currently being manufactured in the USA. A degree of reaction force compensation is achieved by shaping the

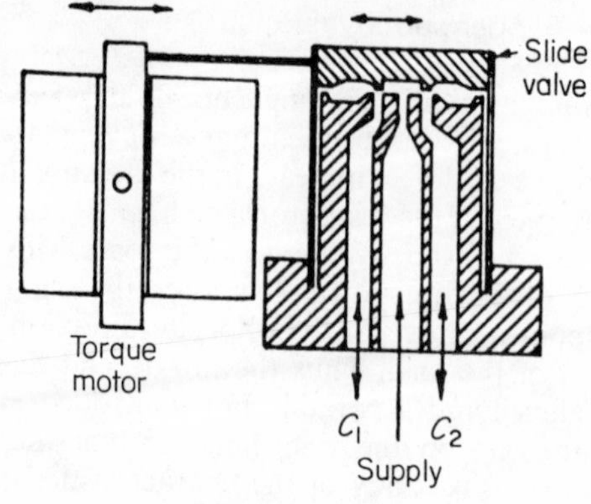

Figure 6.24. Swinging-plate valve (oilgear)

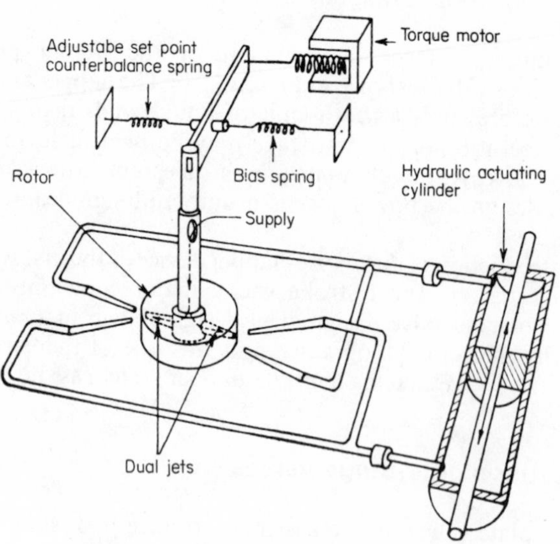

Figure 6.25. Balanced Askania-type valve (GPE controls)

plate profile and the valve itself may be regarded as a return to the D-type design or a two-dimensional version of the valve shown in *Figure 6.13a*.

A method of obtaining radial balance with the Askania-type valve is illustrated in *Figure 6.25*. The oil flow from the supply is divided and fed out from the rotating disc in diametrically opposed directions such that any net radial reaction is minimised. Such valves do not usually have a high-frequency response (i.e. it is much less than 100 Hz) because of the size of the rotor but nevertheless they are adequate for many machine-tool applications.

6.11 Other two-stage valves

In recent years numerous variations in the design of two-stage valves have appeared and each has been said to improve the valve's characteristics generally with respect to some specific application. A typical example of this has been valves capable of retaining high sensitivity whilst incorporating means of dealing with a relatively dirty or contaminated working fluid and further details of these are given below.

6.11.1 Composite double-spool valve

A German design, which may be regarded as a modified or composite double-spool valve shown in *Figure 6.26a*, was used in a Siemens hydraulic autopilot during the 1939–45 war. Earlier designs of composite, double-spool valves are common and each has the advantage of requiring no mechanical linkage or other additional

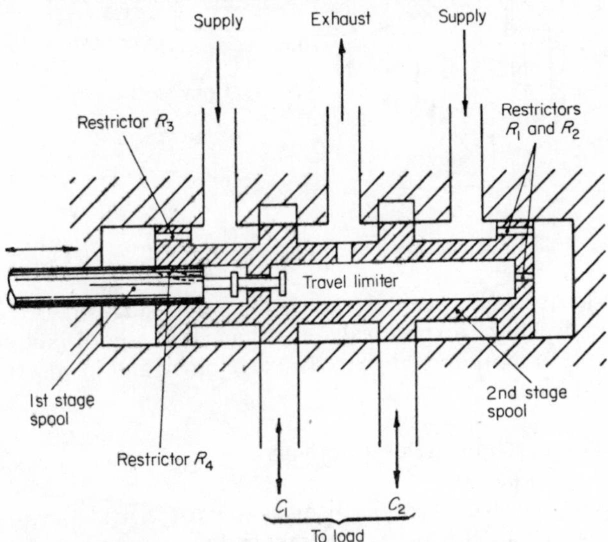

Figure 6.26(a). Composite-spool valve (Siemens)

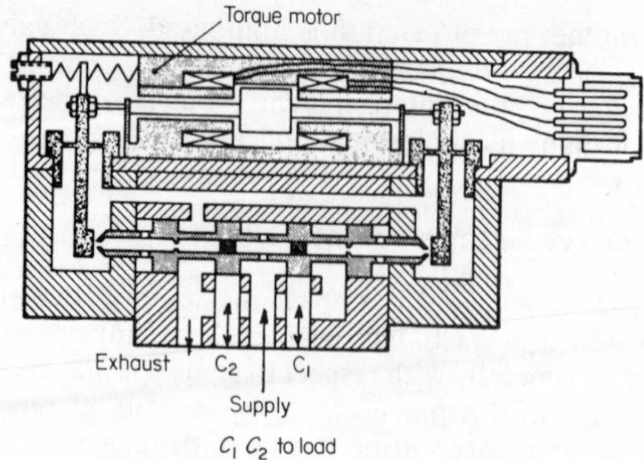

Figure 6.26(b). Composite nozzle-spool valve (Pegasus)

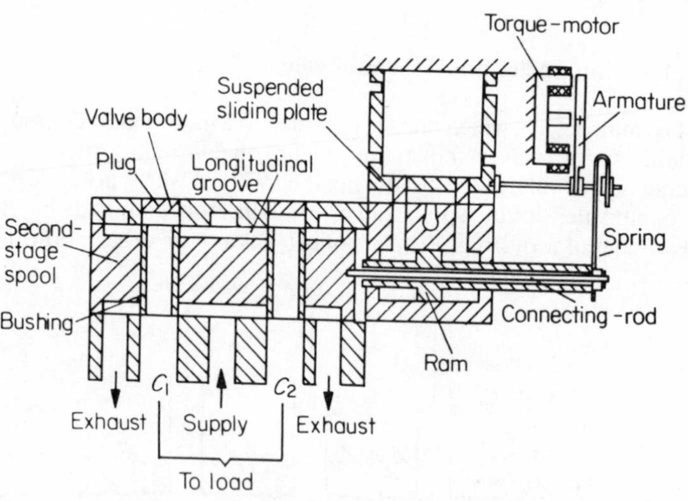

Figure 6.26(c). Two-stage plate spool valve (MIT, DACL)

form of feedback between the first and second stage spools. They often are, in effect, a spool valve in which both spool and sleeve are capable of motion relative to the valve block and in which both sleeve and block are ported.

6.11.2 Composite nozzle-spool valve

It has been suggested[41] that the Siemens valve described above is the forerunner of the later types of composite nozzle-spool valves, an example of which is shown in *Figure 6.26b*. It is of interest to note that the former may be likened to a Wheatstone bridge in which the spool will always seek the position at which the ratio $R_1 R_4/R_2 R_3$ is constant, where R refers to the flow resistance of the orifices so marked in *Figure 6.26a*. The valve shown in *Figure 6.26b*, is similar in operation but two of the resistances are variable as opposed to only one (R_4) in the previous example. These flow restrictions, produced by the presence of the two flappers adjacent to the nozzles in the ends of the spool, vary simultaneously, but as one increases the other decreases (back-to-back operation) thereby increasing the gain of the valve. Two difficulties are often encountered in this design; it requires the use of a special linear torque motor (as opposed to one of rotary design) having a stroke equal to the maximum spool travel and, the sealing diaphragms, isolating the torque motor from the nozzles and through which the flappers pass, often introduce null shift and other spurious and deterimental effects.

An alternative form of this type of modification is produced by Kearfott in the USA, in which the axes of the nozzles, whilst situated in the second stage or main spool, lie along a radius of the spool rather than along its main axis. A saddle-like flapper is used to provide the variable flow restrictors which again operate in a back-to-back manner. A valve of similar design and developed by Eynon[9] has the flapper directly coupled to the torque motor shaft and rotates about an axis perpendicular to that of the second-stage spool (see *Figure 6.27c*).

6.11.3 Plate-spool valve

An unusual two-stage valve insofar as it has a swinging-plate first-plate stage controlling the position of a small ram which in turn is coupled to a spool valve having 'hole and plug porting' (see Reference 40) is shown in *Figure 6.26c*. It was developed at the MIT for a ten horsepower application. Feedback from the second to the first stage is provided by the spring on the right-hand side of the diagram.

6.11.4 Hydraulic feedback

The term 'hydraulic feedback' is generally used when describing the type of valve shown in *Figure 6.27a*, in which the balance of the flow restrictor ratios, after an initial disturbance has occurred is obtained by back-to-back variation in the flow restrictors (orifices or tapered slots) upstream of the nozzles. This again may be considered as an

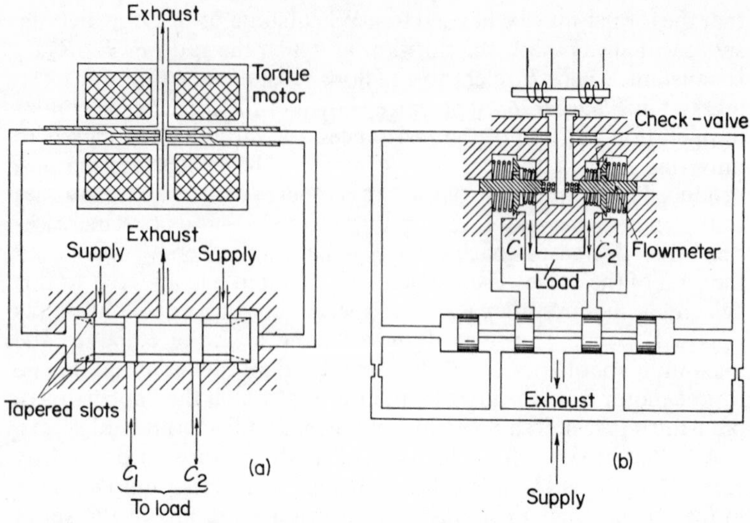

Figure 6.27. (a) Two-stage valve with hydraulic feedback (Armstrong Whitworth Equipment); (b) Two-stage valve with flow rate feedback (Pesco Products)

extension of the German composite spool design principle and valves employing this principle are now being manufactured in both Britain and the USA.

6.11.5 Flow rate feedback

In some systems it is desirable that a feedback signal directly proportional to load velocity or flow rate is available rather than that produced by the main spool displacement (it is only when the load velocity is constant that these latter two quantities have a linear relationship). A recent design, the production of which is now believed to have ceased, is shown in *Figure 6.27b*, and incorporates a simple type of flow-meter in the form of poppet valves. Despite the use of poppets having low-pressure drop characteristics it would seem that the large and sudden variation in the discharge coefficient with poppet travel of such devices would introduce non-linear effects such that little if any advantage be gained.

An additional advantage[42] claimed for this type of valve is that the manufacturing tolerance on the spool lands can be increased by approximately a factor of ten and this should naturally make it extremely popular from the production point of view.

6.11.6 Flapper reaction forces

A field which appears to have received scant if any attention in the USA but to which a significant contribution has been made by Eynon[9] is that concerning flow reaction forces on the flapper of two-stage valves. Considerable changes in both magnitude and sign can be produced in the force required to operate the flapper by re-shaping its surface adjacent to the nozzles.

A valve design incorporating the technique is shown in *Figure 6.27c* and its mode of operation is almost identical to the double nozzle-flapper: spool design, the main difference being that the flapper is split and rotates.

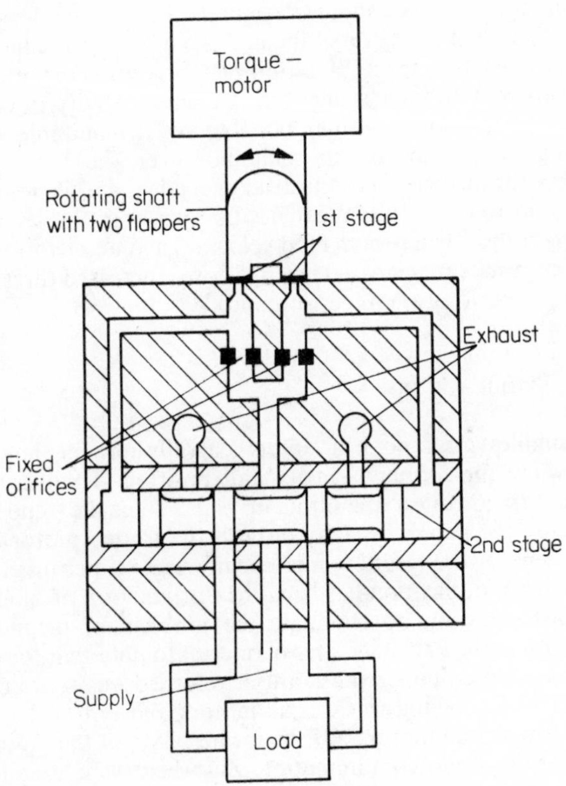

Figure 6.27(c). RAE two stage valve

6.12 Contamination

In the earlier designs of servo-valves, which are mostly for missile applications, the object was to produce a valve of light weight with a good frequency response (above 100 Hz if possible) and a high sensitivity. This was partly achieved by the use of light weight first stages and small torque motors. Considerable trouble arose from the effects of working fluid contamination, and this resulted in the modification of some basic valve designs to reduce the first stage susceptibility to contamination, and also to the development of new designs with a higher level of contamination acceptance.

6.12.1 Effects of contamination

Contamination has two main effects; one is to reduce the reliability of the valve by clogging or silting and the other is to reduce the life by increasing the wear rate of the fine metering edges of the spool and liner. Although an improvement in the valve's ability to operate in relatively dirty oil can often be obtained by suitable re-design (resulting in an increase in reliability over the longer periods necessary for aircraft and industrial use) adequate filtration is still essential to reduce wear. Wear occurs mainly in the second-stage metering orifices where high fluid velocities and accelerations occur. This wear results in increased leakage flows, increased threshold and a general deterioration in performance.

6.12.2 Torque motors

The smallest clearance in a valve is generally between the spool and sleeve when the former is in its null position. To overcome the increases stiction due to the silting up of this stage the trend has been to increase the power capacity of both the torque motor and first stage whenever the resulting increase in weight is permissible.

A further modification has been the introduction of 'stagnant' or 'dry' torque motors to eliminate the problem of magnetic contamination in the first stage. It is of interest to note that for one type of double nozzle-flapper valve, a user returned due to failure, 9.6% of those valves having wet torque motors and only 3.2% of those having dry torque motors. Of the former, 38% of the failures were due to torque motor contamination while there were no failures due to this cause with the dry torque motor type. However, only 0.5% of the failures of the wet torque motor type were attributed to a damaged flapper, while 10.4% of the dry type had failed from this cause. It is possible that this increase in flapper damage was caused by the presence of the sealing diaphragm isolating the torque motor from the working fluid. For an excellent presentation of the design and other details, the reader is referred to Reference 43.

6.12.3 Filtration

Although in nozzle-flapper valves the diameter of the nozzle is about 500 microns and the diameter of the fixed orifice is about 140 microns, the clearance between the nozzle and the flapper is only 25 to 50 microns and particles larger than this will cause jamming of the flapper, silting and consequent clogging of the nozzle, or a transient movement of the load until the particle is cleared. To prevent this integral filters are usually fitted immediately upstream of each orifice and nozzle. In double nozzle-flapper valves unbalance can occur due to filter blockage at one of the nozzles, and to avoid this one valve uses a common filter for both nozzles. This filter, however, cannot be placed immediately upstream of the nozzles and thus does not remove contamination which might be left in the passageways during assembly. Other filtration techniques have been used such as flushing the outside of the filter with the second-stage flow, and the use of filter inlets arranged at 90° to the main flow direction so that the momentum of the larger particles helps to prevent them from entering and subsequently clogging the filter.

6.12.4 Nozzle and flapper arrangements

The advantages which are claimed for the single-nozzle (see *Figure 6.23b*) over the double-nozzle type valve (see *Figure 6.23c*) are that since only one metering orifice is used, the critical dimensions can be larger for the same quiescent flow. In addition the flapper is able to move much further away from the nozzle if it is necessary to clear particles.

6.12.5 Chip shearing

Some manufacturers include a 'chip shearing' action in the valve design so that if they become jammed by a chip a large force, capable of shearing the chip, is applied to the spool or sleeve. It has been pointed out[38] that this self-clearing action is very seldom a desirable feature since it has two overiding disadvantages; the first being that the resulting transient disturbance of the system whilst the chip is being sheared, followed by the restoring action, could result in an unacceptable displacement of the output , e.g. serious damage in machine-tool applications. The second disadvantage is that if it were a steel chip it is probable that the resulting damage and chipping of the hardened, steel, metering edges being thus used as shearing blades would ruin the valve for critical applications.

6.13 Recent valve designs

A valve which was initially designed for a higher level of contamination acceptance is the two-stage Askania: spool design shown in *Figure 6.28*. The smallest clearance is the Askania-type nozzle

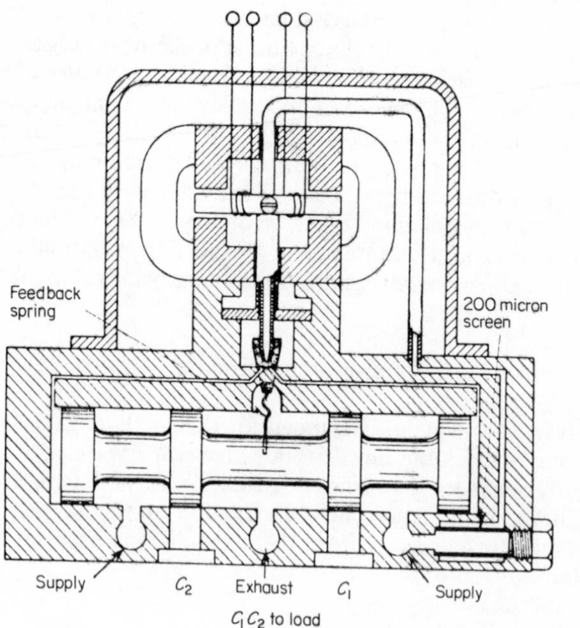

Figure 6.28. Two-stage Askania: spool valve (Raymond Atchley)

diameter which is about 250 microns which is protected by a 200 micron screen. It is claimed that long, needle-like particles, which would quickly clog the usual flapper-nozzle designs and which are a fairly common type of contaminant, can be passed by the first stage of this valve and cause little, if any, trouble.

6.14 Valve testing

In addition to the previously mentioned difficulty of distinguishing one type of valve design from another, it is also sometimes difficult to assess and compare performances of valves. This is particularly so since, unlike pump manufacturers, few valve manufacturers use the same methods of testing and one is often faced with a multitude of test results each of which has been measured under different values of valve pressure drop, input signal and load conditions.

It is not desirable to stipulate rigidly valve testing specifications, but of the many which have been issued the specification given by Dowty Rotol Ltd[43] is probably the most useful and comprehensive. In addition to covering terminology and definitions, electrical and hydraulic design data and performance specifications, it suggests a specific test procedure and the various test curves and results which should be obtained. These are:

1. Null. The input signal at which the valve is in its null position.
2. Polarity. (Confirming the functions of the various ports).
3. Null pressure gain characteristic.
4. Quiescent flow.
5. Flow linearity, symmetry, gain. (A plot of no-load flow/input).
6. Hysteresis. (Width of hysteresis loop of no-load flow/input curves).
7. Resolution. (Increment of input required to produce a change in valve output).
8. Flow/load pressure characteristic.
9. Proof pressure. (Leakage test at 1.5 times rated supply pressure).
10. Null shift with supply pressure.
11. Null shift with return pressure.
12. Null shift with quiescent current.
13. Null shift with temperature.
14. Null shift with acceleration (effect of acceleration on valve performance).
15. Amplitude ratio.
16. Phase lag.

6.15 Conclusions

In the absence of a comprehensive series of comparative test results for each design of valve it is virtually impossible to suggest, in general terms, which if any valve design is the best. In addition, although a detailed specification of a required performance will probably help by eliminating designs which are obviously unsuitable, there will in all probability be several others which can meet it. Alternatively, there may be several which fail on only one or two items and these items would probably be different for the different designs. An interesting example of a condition similar to this is illustrated in *Figure 6.29*, where the null shifts of several valves are shown when the valves are subjected to a temperature cycle of 38° to 93° to 38 °F.

The percentage null shift has been plotted against the frequency at which the response of each valve changes by 3 dB and illustrates that low-null shift and high-frequency response requirements usually conflict. To some extent this is to be expected since the performance of a valve which has been accurately tuned to give a high frequency response will be more susceptible to deterioration caused by external disturbing factors such as vibration etc. The figure also shows that, in general, mechanical feedback (via levers, springs, etc.) designs are preferable to the use of a spring restrained, second-stage spool design when a minimum value of null shift at any specified frequency is required.

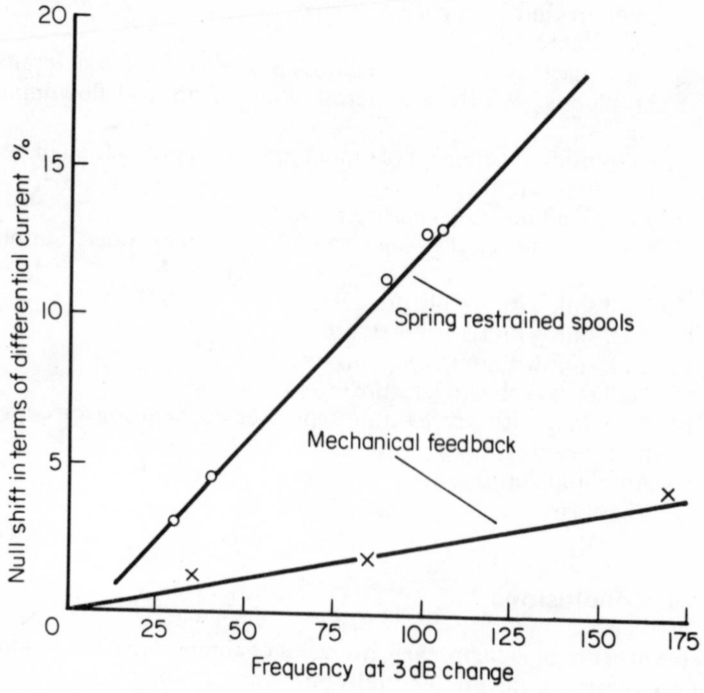

Figure 6.29. Variation of null shift with performance

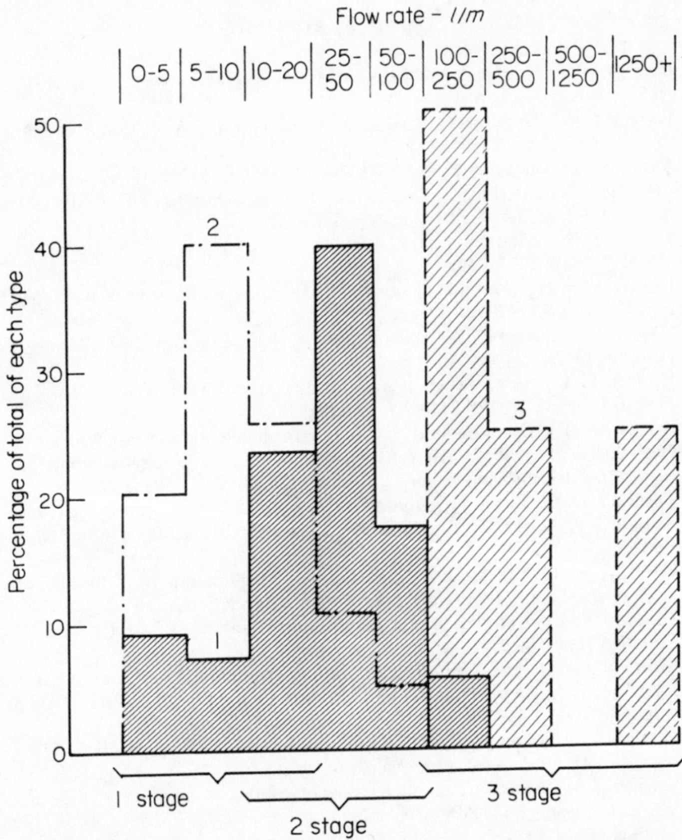

Figure 6.30. Usual flow rate ranges of single stage, two-stage and three-stage valves

Although single-stage valves can generally provide adequate performance for small maximum flow rates and powers, two-stage valves become popular as maximum flow rates increase. *Figure 6.30* illustrates this point and is a histogram of valves (grouped together according to the number of stages) manufactured in the United States on a basis of flow rate. The number in each flow rate range is given as a percentage of the total number of valves having one, 100 litre/min. In addition, single-stage valves are generally used for flow rates less than 10 litre/min and three-stage valves are commonly used for flow rates in excess of 100 litre/min. A certain amount of overlap does, however, exist and it would not be wise to regard the above limits as being hard and fast boundaries.

REFERENCES

1. Beyer, C., 'On balancing the valves of steam engines', *Proc. I.Mech.E.* p. 189 (1857).
2. Wymer, F. W., 'Description of a double slide expansion valve for marine engines', *Proc. I.Mech.E.* p. 58 (1856).
3. Ansaloni, A., 'The Eiffel Tower lifts', *Proc. I.Mech.E.* p. 350 (1889).
4. Brading, T. and Parkes, G., 'Some industrial applications of hydraulic servos', *Proc. Conf. Hyd. Servos., I.Mech.E.* p. 33 (1953).
5. Smith, E. S., 'Automatic regulators, their theory and applications', *Trans. A.S.M.E.* **58,** p. 291 (1936).
6. Douch, E. J. H., 'The use of servos in the army in the past war', *JIEE* **94,** Part IIA, p. 177 (1947).
7. Garde, A., *Automatic and manual control,* 1st edition, p. 503, Butterworth, London (1952).
8. Harpur, N. F., 'Some design considerations of hydraulic servos of jack type', *Proc. Conf. Hyd. Servos. I.Mech.E.* p. 41 (1953).
9. Eynon, G. T., 'Developments in high-performance electro-mechanical servomechanisms at the Royal Aircraft Establishment', *Proc. Sym. Recent mechanical engineering developments in automatic control, I.Mech.E.* p. 93 (1960).
10. Turnbull, D. E., 'Piston-type control valves', Ph.D. Thesis, Cambridge (1956).
11. Khoklov, B. A., 'Electrohydraulically controlled drives', 1st edition, Moscow (1964).
12. Conway, H. G. and Collinson, E. G., 'An introduction to hydraulic servomechanism theory', *Proc. Conf. Hyd. Servos. I.Mech.E.* p. 1 (1953).
13. Kinney, W. L. *et al.*, 'Hydraulic servo control valves', *W.A.D.C., T.R. 55-29,* Parts I–VII (1958).
14. Montgomery, J. and Lichtarowicz, A., 'Assymmetrical lap and other nonlinearities in valve-controlled hydraulic actuators', *Proc. I.Mech.E.* **183,** pt. 1 (1968/9).
15. Houlobek, F., 'Comparative rigidities of certain valve controlled hydraulic servos', *RAE Tech. Note, M.E. 129* (July 1952).
16. Shearer, J. L., 'Resistance characteristics of control orifices', *Proc. Sym. Recent mechanical engineering development in automatic control, I.Mech.E.* p. 35 (1960).
17. MacLellan, G. D. S. *et al.*, 'Flow characteristics of piston-type control valves', *Proc. Sym. Recent mechanical engineering development in automatic control, I.Mech.E.* p. 13 (1960).
18. Viersma, T. J., *Investigations into the accuracy of hydraulic servomotors,* Phillips Research Report, 16, p. 507–596 (1961).
19. McCloy, D. and Beck, A., 'Some cavitation effects in spool valve orifices', *Proc. I.Mech.E.* **182,** Part 1 (1967–68).
20. McCloy, D. and Beck, A., 'Flow hysteresis in spool valves', *BHRA, SP 989* (January 1969).
21. Clarke, R. N., 'Compensation of steady state flow forces in spool type hydraulic valves', *Trans. A.S.M.E.,* **79,** p. 1784 (1957).
22. Carrington, J. E., 'Developments in electro-hydraulic valves', *Hyd. Pneu. Power* **12,** 135, p. 154 (March 1966).
23. McCloy, D. and Martin, H. R., 'Some effects of cavitation and flow forces in the electrohydraulic servomechanism', *Proc. I.Mech.E.* **178,** Part I, p. 539–58 (1964).
24. Swain, A. E., *Time lags, amplitude variation and force reaction of hydraulic relays,* Nash and Thompson, T.N. 98 (1942).
25. Sweeney, D. C., *Out of balance reactions in hydraulic, piston-type control valves,* Ph.D. Thesis, Birmingham (1949).

26. Lee, S-Y. and Blackburn, J. F., 'Contributions to hydraulic control', *Trans. A.S.M.E.* **74,** pp. 1005, 1013 (1952).
27. von Mises, R., 'Berechnung von ausfluss-und verberfallzahlen', *Z.V.D.I.*, **61,** p. 494 (1917).
28. Feng, T-Y., 'Steady state axial flow forces on pneumatic spool-type control valves', *ASME Paper No. 57-A-129* (Dec. 1957)
29. British Patent No. 843,832 (August 1960).
30. Alcock, J. F., *Proc. I.Mech.E.*, **158,** p. 203/5 (1947).
31. Royle, J. K., contribution to paper 'Hydraulic servos' by R. Hadekel, *Proc. Conf. Hyd. Servos. I.Mech.E.* p. 65–6 (1953).
32. Stringer, J. E. C., 'Hydraulic lock: another explanation'. *Engineering* (April 25, 1952).
33. Whiteman, K. J., 'Hydraulic lock at high pressures', *The Engineer* **203,** No. 5281, pp. 554–7 (L2 April 1957).
34. Weber, J. E., *Lateral forces on hydraulic pistons caused by axial leakage flow*, S.M. Thesis, M.I.T., Cambridge, Mass., U.S.A., p. 30–4 (1951).
35. Faisandier, J., 'Rotary servo-valve for aircraft applications', *App. Hyd. Pneu*, No. 117 (Feb. 1959).
36. Shute, N. A. and Turnbull, D. E., 'Some trends in electro-hydraulic servo-valve design', *Proc. Conf. Hyd. Power Trans. I.Mech.E.* (1961).
37. Hadekel, R., 'Hydraulic servos', *Proc. Conf. Hyd. Servos, I.Mech.E.* (1953).
38. Blackburn, J. F. *et al.*, *Fluid power control*, 1st edition, J. Wiley and Sons, London (1960).
39. Besser, R. I., 'New concepts of split spool valving', *ASME Paper No. 69-DE-39* (May 1969).
40. Lee, S-Y., 'Contribution to hydraulic control', *Trans. A.S.M.E.* **76,** p. 905 (1954).
41. Gibson, J. E. and Tuteur, F. B., *Control system components*, 1st edition, p. 425, McGraw-Hill, London (1958).
42. Bahniuk, E. and Lee, S-Y., 'The design and analysis of a servo-valve with flow feedback', *ASME Paper 59-IRD-3* (March 1959).
43. Merritt, H. E., *Hydraulic Control Systems, 1st ed*, John Wiley & Sons Inc. (1967).
44. *Specification Standards for electrohydraulic flow control servovalves*, Dowty Rotol Ltd., Gloucester (1960).

Chapter 7

System Response

Although there is a very large number of designs of hydraulic servo systems, it is often convenient and possible to reduce the analysis of their response to that of the relatively simple device shown in *Figure 7.1* in which the behaviour of all components except the valve and jack has been assumed to be ideal in that they have no time delays

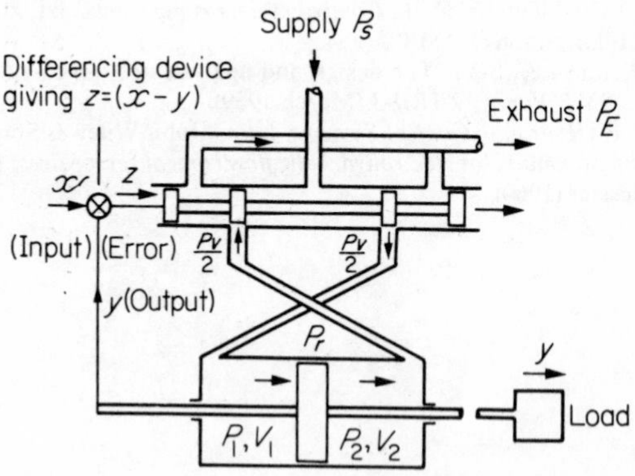

Figure 7.1. Arrangement of system to be studied

and non-linear effects. In practice, this is generally far from the truth, particularly if the system is required to respond to a high frequency input signal. Nevertheless this relatively simple approach is of value since it focusses attention on the behaviour of the hydraulic parts of the system and it is these with which this chapter is primarily concerned.

7.1 The general response equation

7.1.1 Early work

The earliest published work on the loaded hydraulic, position control system appears to be that of Coombes[1] who derived the response equation for an inertial load. Because no analytical solution was available for the sinusoidal response he then went on to derive the waveform of the input required to produce a sinusoidal output. From the results obtained it was possible to deduce that an inertial load had little effect on the dynamic behaviour of the relay provided no more than half the pressure drop of the entire system occurred across the load. Similar work was done by Grosser[2] in connection with step responses but the presence of relief valves at either end of the actuator limited the application of his results.

A few years later, Gold, Otto and Ransom[3] obtained some experimental results and attempted to correlate them with a linearised response equation for both step and sinusoidal inputs. Only an inertial load was considered and the validity of the approximations used in the linearisation is open to question. The importance of a knowledge of the nature of the load was emphasised by Lewis[4] who examined the effects of valve reaction forces and supply pressure on system response, and a few theoretical results concerning step responses illustrated the importance of reaction forces. At the same time this work was extended by Heinz[5] who showed that when reaction forces were present the maximum power from the system (see section 7.7) was obtained when less than one-third of the available pressure drop occurred across the valve which was the condition previously derived by Conway and Collinson[6].

The linearised method used by Gold *et al*[7,8,9] was adopted by Mucha[10] to derive step responses for constant valve openings but as analytical solutions to the exact equations exist it is of little interest. In a study of machine tool problems Stallard[11] managed to account for fluid compressibility by considering only output velocity. In this way he avoided the third-order equation which is usually obtained when this subject is studied (see section 7.6.2) and a similar approach by Royle[12] is described in section 7.6.2.

An exact but graphical solution for an inertial load with a sinusoidal input was obtained by Noton and Turnbull[13] for the condition of no feedback but the work was not extended since they were only interested in valve instability caused by valve reaction forces. Royle[14] obtained a series solution for the inertial load response and, with the aid of a differential analyser, also studied combinations of inertial and viscous loads. His results are discussed in more detail in section 7.6.1. Some work has also been done on the response of the hydrostatic transmission system consisting of a pump and motor and a description of this has been given in Chapter 5, section 5.6.1.

7.1.2 Derivation of the general response equation

Considering the system shown in *Figure 7.1* it will be assumed that

(a) the areas of the two control ports are always equal and proportional to the valve travel, z, which implies zero lap and rectangular ports, and

(b) that the fluid, which is generally oil, is incompressible although the effect of compressibility will be studied in detail later in section 7.6.2.

If it is assumed that the exhaust or drain pressure is zero then

$$P_s = P_v + P_r \tag{7.1}$$

where P_v is the *total* pressure drop across the valve ports and is assumed to be equally divided between the two controlling orifices. This in turn is taken to imply equal flow rates through each orifice or control port and provided the discharge coefficients do not differ greatly the error in practice is small.

The error or spool travel z is given by

$$z = (x - y) \tag{7.2}$$

that is the difference between the input x and the output y, and the flow rate Q through each port is given by

$$Q = kz\sqrt{P_v} \tag{7.3}$$

where $k = n'wC_d/\sqrt{\rho}$

n' is the number of ports;
w, the circumferential width of one of the ports;
C_d, the discharge coefficient;
ρ, the fluid density.

If $f(L)$ is the force required to move the load then

$$f(L) = P_r A_r \tag{7.4}$$

and combining equations 7.1 to 7.4 and writing $\mathrm{d}y/\mathrm{d}t = Q/A_r$ then gives

$$\frac{\mathrm{d}y}{\mathrm{d}t} = \frac{1}{T}(x - y)\sqrt{\left(1 - \frac{f(L)}{P_s A_r}\right)} \tag{7.5}$$

where T is defined as the time constant and is given by

$$T = \frac{A_r\sqrt{\rho}}{n'wC_d\sqrt{P_s}} \tag{7.6}$$

Equation 7.5 is the governing or response equation of the system and it may be thought of as:

Flow rate $\propto$ port area. $\sqrt{}$(Pressure drop across valve ports)

or as

Velocity $\propto$ error. $\sqrt{}$(Supply pressure–load pressure drop).

7.2 The step response

Dividing both sides of equation 7.5 by x, multiplying by T, writing $t/T = \bar{t}$ where $\bar{t}$ is a non-dimensional form of time and writing $y/x = \bar{y}$ where $\bar{y}$ is the output ratio then gives

$$\frac{d\bar{y}}{d\bar{t}} = (1-\bar{y})\sqrt{\left(1-\frac{f(L)}{P_s A_r}\right)}$$

and denoting $d\bar{y}/d\bar{t}$ by $\dot{\bar{y}}$ then gives

$$\dot{\bar{y}} = (1-\bar{y})\sqrt{(1-f(L)/P_s A_r)} \tag{7.7}$$

7.2.1 The 'no-load' response

If there is no load on the ram then $f(L) = 0$ and the response equation becomes

$$\dot{\bar{y}} = (1-\bar{y}) \tag{7.8}$$

and if the input x, consists of a sudden step of magnitude X the solution of equation is

$$y/X = 1-e^{-t/T} \tag{7.9}$$

This is a linear, first order or exponential response and is shown in *Figure 7.2*. Under these conditions it should be noted that equation 7.8 may be written in the form

$$\bar{y}+\dot{\bar{y}} = 1$$

or

$$\bar{y}(1+TD) = 1$$

where D is the operator d/dt and $d\bar{t} = dt/T$. As a result the transfer function of the system which is defined as the ratio of the output to the input written in terms of the 'D' operator and in this case equal to $\bar{y}\,(=y/X)$ is given by

$$\frac{y}{X}=\frac{1}{(1+TD)} \tag{7.10}$$

which is again recognised as a linear first order response having a single time constant T.

7.2.2 The system time constant

The time constant T of the system is given by equation 7.6 as

$$T = A_r\sqrt{\rho}/n'wC_d\sqrt{P_s}$$

and the origin of its name follows from the fact that the combined dimensions of the group of parameters on the right-hand side of the equation is simply that of time. The magnitude of a system's time

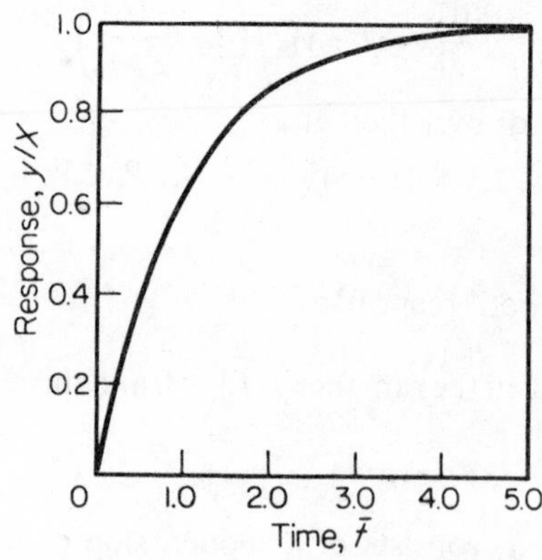

Figure 7.2. Step response of the system with no-load

constant is of great importance since it governs the speed at which the system can respond to an input signal. For example, referring to the response shown in *Figure 7.2* it can be seen that the output motion is nearly complete after a non-dimensional time interval, $\bar{t}$, of approximately 4. The actual time, t, which has elapsed is given by $t=\bar{t}T=4T$ so that if the time constant, T, of a system were halved then the system output would travel twice as quickly as it did before the alteration were made. As a result designers are always searching for methods of reducing system time constants in an attempt to increase the speed of response.

It should be noted that after a time interval of unity, i.e., $\bar{t} = 1$ or $t = T$, the first order system response can be shown to have reached approximately 63% of its final position and if the initial value of the velocity had been maintained throughout the entire response the motion would have been completed in this time of $t = T$. This is sometimes used as a definition of a system's time constant.

7.2.3 Response time

The non-dimensional time taken by a simple first-order system to complete 98% of its final travel is 3.912 and as stated above it may therefore be said that motion is virtually complete after a period of approximately four times the time constant. To assess the speed of the step response of a loaded system the time taken to move 98% of the total travel and defined as the response time $\bar{t}_r$ will be used and compared with the approximate 'no-load' value of the response time of 4. This somewhat arbitrary method of comparison is necessary since in theory many such systems take an infinite time to reach their final output position and this time, at which motion is complete, cannot therefore be used for comparison purposes.

With a sinusoidal input to the system it is possible to make a direct comparison between the responses with and without the load in terms of the amplitude and phase angle of the output position as described in section 7.5. This procedure becomes a little difficult under conditions when the output wave form ceases to be sinusoidal but it will be seen later that some allowance for this effect can be made.

7.2.4 Step responses with various loads

A detailed study of the effect of various loads on the step response of the system is given in References 15 and 16. Since they contain a considerable amount of detailed mathematical analysis only a brief review of the main results will be given. *Table 7.1* gives the types of load considered, the corresponding values of $f(L)$ and the coefficient of $\bar{y}$ or of its time derivatives present in the square root term of the response equation.

Two typical sets of response curves are shown in *Figure 7.3* for various viscous and inertial loads on the ram and all the results are summarised in *Figure 7.4* by using the response time defined above.

Table 7.1 DETAILS OF RESPONSE EQUATIONS FOR STEP INPUTS

Type of Load	$f(L)$	*Coefficient of* $\bar{y}$, $\dot{\bar{y}}$, *or* $\ddot{\bar{y}}$
No load	0	0
Constant	F_c	$L_f = F_c/P_s A_r$ (Note: no y term)
Spring	sy	$L_s = SX/P_s A_r$
Viscous	$\mu' \, dy/dt$	$L_v = X/P_s A_r T$
Orifice	$r(dy/dt)^2$	$L_r = rX^2/P_s A_r T^2$
Inertial	$M \, d^2y/dt^2$	$L_m = MX/P_s A_r T^2$

F_c, constant force opposing ram motion.
s, rate of a spring opposing ram motion.
μ', coefficient proportional to the viscous load on the ram.
r, coefficient proportional to the orifice type load on the ram.
M, inertial load on the ram.

The effects of a composite load of a spring and a viscous load have also been derived and again the results are presented in the response time form in *Figure 7.5*.

When an inertial load is present it was found that the response equation could not be integrated to give $\bar{y}$ as a function of $\bar{t}$ but the relation between the displacement ($\bar{y}$) and its velocity ($\dot{\bar{y}}$) could be obtained. If these two quantities are plotted against each other then the resulting curve is usually referred to as the trajectory in the 'phase plane' (see section 7.3) and two trajectories for a combined spring and an inertial load are shown in *Figure 7.6*.

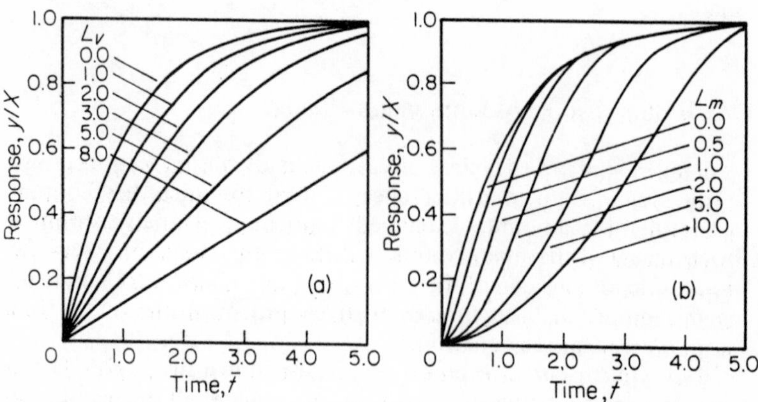

Figure 7.3. *Step responses.* (left) *Viscous loads;* (right) *Inertial loads*

7.2.5 Summary of results for step responses

The presence of a constant load on the output of the system does not make the system non-linear but merely increases the effective time constant and therefore reduces the speed of response. There is no need to consider individual response curves (y/X against t) corresponding to various values of L_f for they will be identical with the curve shown in *Figure 7.2* if the time parameter is replaced by $t\sqrt{(1-L_f)}/T$.

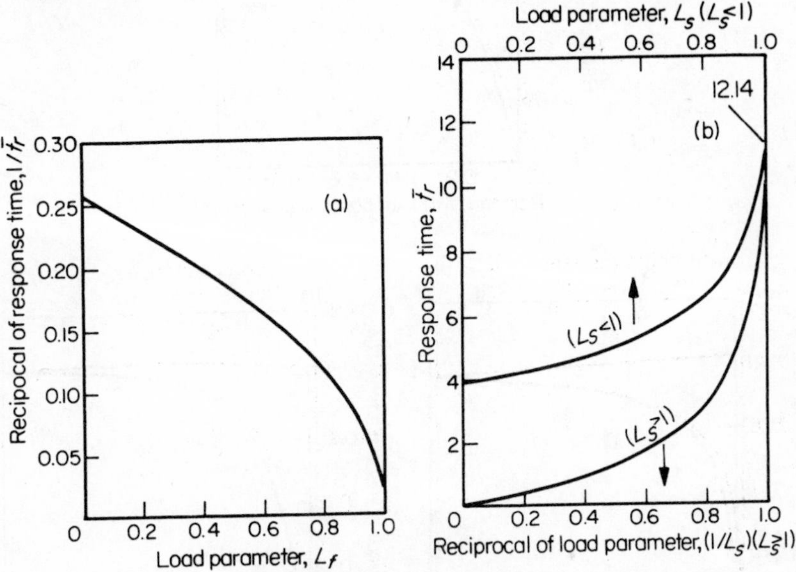

Figure 7.4. Variation of response times with magnitudes of loads. (a) Constant load; (b) Spring load.

The maximum response time when a spring load is present is 12.14 and occurs when the load parameter L_s approaches unity. The response equation for this condition is of particular interest for it is identical with that of the system when the valve has circular ports (see section 7.6.5), the input to the system is small compared with the port diameter and there is no load on the ram. The values of the response time, $\bar{t}_r$, given for values of L_s greater than unity are probably of academic interest only for it is unlikely that any practical system would be designed to operate under such conditions.

Both orifice and viscous loads were found to have a large effect on the initial velocity of the response and, unlike the spring load response, the initial velocity ($\dot{\bar{y}}$ when $\bar{y} = 0$) is always less than unity.

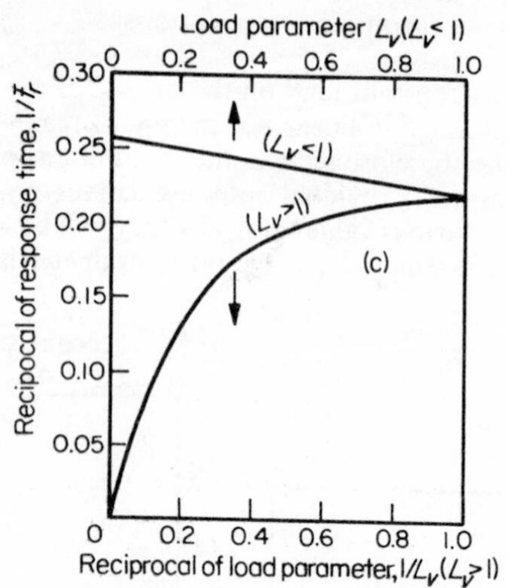

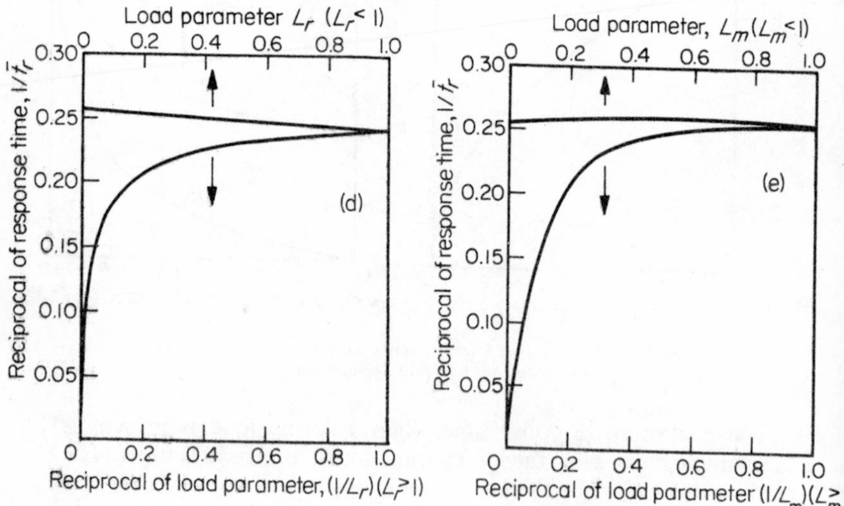

Figure 7.4 (continued)
(c) Viscous load; (d) Orifice load; (e) Inertial load

This is to be expected from the nature of the load since it will limit the maximum output velocity.

It is surprising how large a value of the inertial load parameter L_m is required to produce a significant increase in the response time (*Figure 7.4e*) and it should be noted that despite the rate at which the responses (*Figure 7.3b*) change their shape with increasing values of L_m there are no appreciable effects on the latter part of the response.

Although an analytical expression can be obtained for the response when a combination of both spring and viscous loads are present its complexity suggests the need for speedier if less accurate means of solving response equations than the derivation of exact analytical solutions. When inertial and spring loads are present,

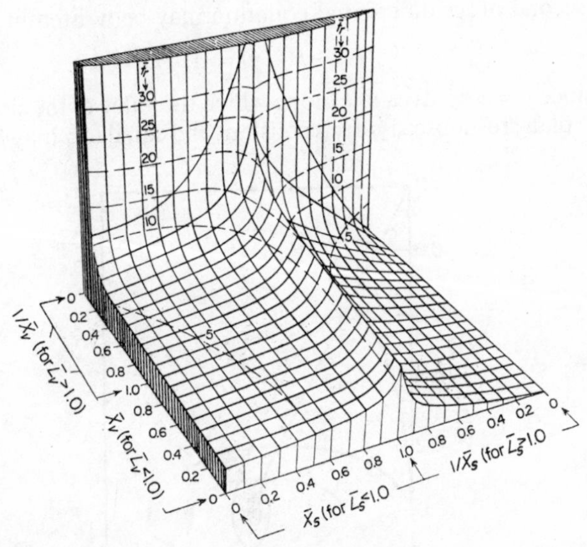

Figure 7.5. Variation of response time for combined spring and viscous loads

however, a reasonably simple, analytical relationship between the velocity and displacement of the system may be obtained. This enables trajectories in the phase plane to be drawn and the usual response curves to be obtained. An oscillatory response can occur if the spring load parameter (L_s) exceeds unity, and two examples illustrating this have been given. The frequency of the oscillations is only a little less than that of a simple mass and spring system.

7.3 The phase plane

The phase plane is a powerful tool for solving non-linear second order differential equations and although it may be tedious to derive solutions for certain conditions it does enable a fairly clear picture of the response to be obtained. Generally the phase plane is shown as a plot of the velocity and the magnitude of the error of a system but for the hydraulic relay with a step input it is convenient to use the velocity $\dot{\bar{y}}$ and magnitude $\bar{y}$ of the output noting that the latter is equal to unity minus the error, since $z = (x-y)$ and $\bar{y} = (1-\bar{z})$.

7.3.1 The use of the phase plane

Any second order differential equation may be written in the form

$$\ddot{y}+f(\dot{y}, y) = 0$$

and since $\ddot{y} = \dot{y}\,d\dot{y}/dy$, a line along which the value of the slope of the phase plane response, $d\dot{y}/dy$, is constant and equal to α may be drawn

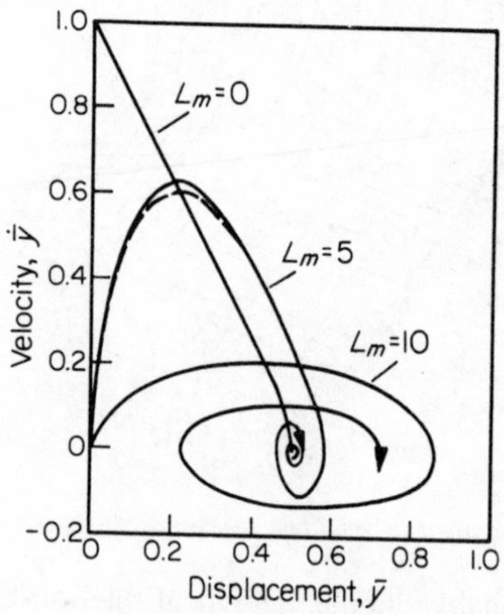

Figure 7.6. Phase plane trajectories of step responses with combined spring and inertial load

by substituting $d\dot{y}/dy = \alpha$ into the original equation and plotting the resulting curve. In other words replacing $\ddot{y}$ by $\dot{y}\alpha_0$ gives

$$\dot{y}\alpha_0+f(\dot{y}, y) = 0$$

and the curve given by this equation may be drawn in the $\dot{y}/y$ plane and every trajectory must cross it with a slope α_0. Such a curve is called an *isocline* and it relates the velocity, $\dot{y}$, to the displacement y, for any given value of α. This process may be repeated for a range of values of α, i.e. α_0, α_1, α_2, etc., as shown in *Figure* 7.7. With the initial conditions of the response say, $y = y_0$ and $\dot{y} = \dot{y}_0$ given, the

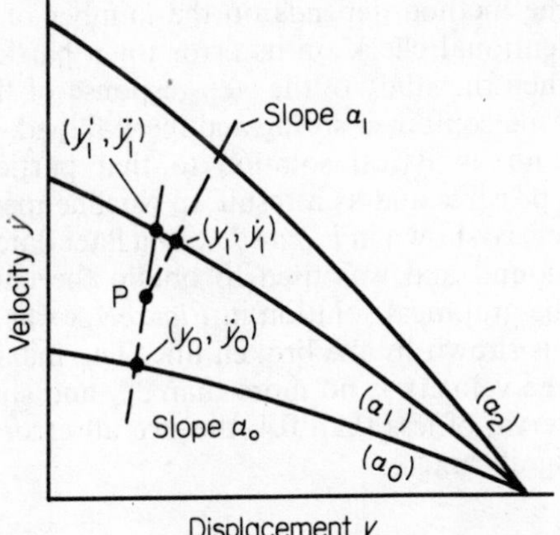

Figure 7.7. Construction of response in the phase plane
Isoclines —— Response - - -

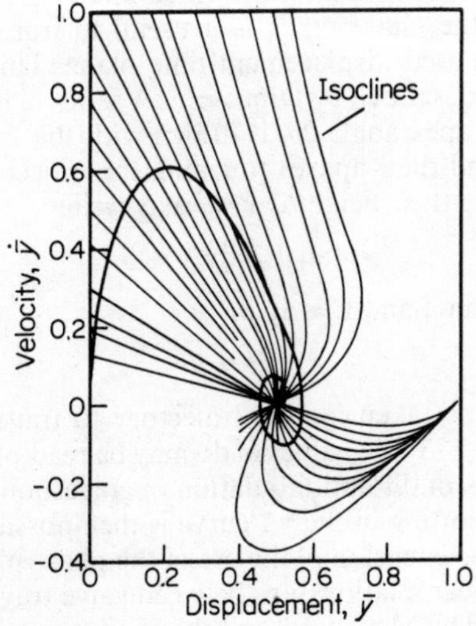

Figure 7.8. Graphical solution of step response for combined spring and inertial loads

starting point of the trajectory is known and a line of slope α_0 is drawn through this point to meet the α_1 isocline at say y_1', $\dot{y}_1'$. A line of slope α_1 is now drawn to intersect the line joining y_0, $\dot{y}_0$ to y_1', $\dot{y}_1'$ at P which is midway between the isoclines for α_0 and α_1 and this will intersect the α_1 isocline at $y_1, \dot{y}_1$ and the two straight lines $y_0, \dot{y}_0$–P–$y_1, \dot{y}_1$ is the approximation to the final trajectory. The accuracy of the method depends on the number of isoclines used and an unintentional check on its error for a particular case was obtained[16] when the study of the step rêsponse of the system was first made for the combined spring and inertial load. It was initially thought that no analytical solution to that particular response equation was possible and as a result an isocline method was used and gave the curve shown in *Figure 7.8*. At a later date the analytical method was found and was used to obtain the curves shown in *Figure 7.6*. The graphical solution in *Figure 7.8* was superimposed on these and is shown by the broken line. The maximum error of the value of the velocity is no more than 3% and since this occurs for a time interval of less than 0.4, the overall error for the entire motion is insignificant.

7.3.2 The transfer of the trajectory to displacement time plane

Although experience enables one to compare responses shown as trajectories in the phase plane is it useful to transfer them to the more generally used displacement/time plane. This is easily done using a method described by Diprose[17a]. A series of isoscles triangles, having a small apex angle 2θ, is drawn with the y axis being used as their base and their apexes touching the trajectory as shown in *Figure 9.9*. Since their height is $\dot{y}$ we may write

$$\tan\theta = \mathrm{d}y/\dot{y} \approx \theta$$

and since θ is small and $\dot{y} = \mathrm{d}y/\mathrm{d}t$

$$\mathrm{d}t \approx \theta$$

which is the time taken for the trajectory to travel a distance $\mathrm{d}y$. By making $\theta = 5.73°$ the values of $\mathrm{d}y$ may be read off corresponding to time intervals of 0.1 and tabulation of corresponding values of y and t and the plotting of the y/t curve is then possible.

An interesting extension of the use of the phase plane to take into account non-linear relationships between valve travel and port area has been derived by McCloy[17b] and shows that considerable changes in the form of the response can occur under these conditions.

7.4 The response of a system with several loads

When several loads, including an inertial one, are present it is unlikely that an analytical solution to the step-response equation exists. Even if one did exist, the time required to obtain it and to calculate points on the response curve would be excessive and a graphical solution by the method of isoclines has much to offer in the way of speed and obtaining an insight into the behaviour of the system.

Suppose the response is required of a system having all the loads previously studied. The response equation when all types of loads previously discussed are present is

$$\dot{\bar{y}} = (1-\bar{y})\sqrt{(1-L_f-L_s\,\bar{y}-L_v\,\dot{\bar{y}}-L_r\,\dot{\bar{y}}^2-L_m\,\ddot{\bar{y}})}$$

and, by redefining the time constant and the load parameters,

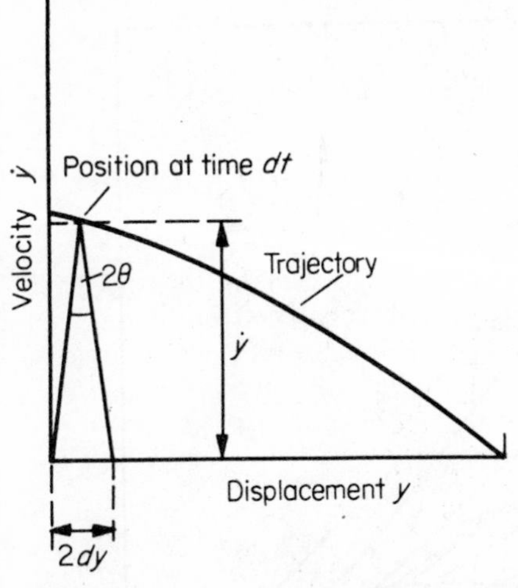

Figure 7.9. Method of transferring trajectory from phase plane to displacement/time place

excluding L_f, to include the terms $\sqrt{\{P_s(1-L_f)}$ and $P_s(1-L_f)\}$ respectively the equation may then be written as

$$\dot{\bar{y}} = (1-\bar{y})\sqrt{(1-L_s\,\bar{y}-L_v\,\dot{\bar{y}}-L_r\,\dot{\bar{y}}^2-L_m\,\ddot{\bar{y}})}$$

The slope of a trajectory in the phase plane is $\mathrm{d}\dot{\bar{y}}/\mathrm{d}\bar{y}$ and the acceleration $\ddot{\bar{y}}$ may be written as before, as $\dot{\bar{y}}\,\mathrm{d}\dot{\bar{y}}/\mathrm{d}\bar{y}$ so that if the slope is constant and equal to α the equation of the isocline for slope α is

$$\dot{\bar{y}} = (1-\bar{y})\sqrt{(1-L_s\,\bar{y}-L_v\,\dot{\bar{y}}-L_r\,\dot{\bar{y}}^2-L_m\,\alpha\dot{\bar{y}})}$$

Solving for $\dot{\bar{y}}$ in terms of $\bar{y}$ gives

$$\dot{\bar{y}} = f(\bar{y}) \pm \sqrt{\left[\{f(\bar{y})\}^2 - \frac{(1-L_s\,\bar{y})(1-\bar{y})^2}{1-L_r(1-\bar{y})^2}\right]}$$

where $$f(\bar{y}) = \left[\frac{(L_v+L_m\,\alpha)(1-\bar{y})}{2\{1-L_r(1-\bar{y})^2\}}\right]$$

It should be noted that as L_m and α are always present as a product, the isoclines may be drawn for various values of (αL_m) and then used for any values of L_m and the corresponding values of α. This point is illustrated by the following numerical example.

Let $L_s = L_v = L_r = 1$. Then,

$$\dot{\bar{y}} = f_1(\bar{y}) \pm \sqrt{\left[\{f_1(\bar{y})\}^2 - \frac{(1-\bar{y})^3}{(2\bar{y}-\bar{y}^2)}\right]}$$

where $f_1(\bar{y}) = f(\bar{y})|_{L_s = L_v = L_r = 1}$.

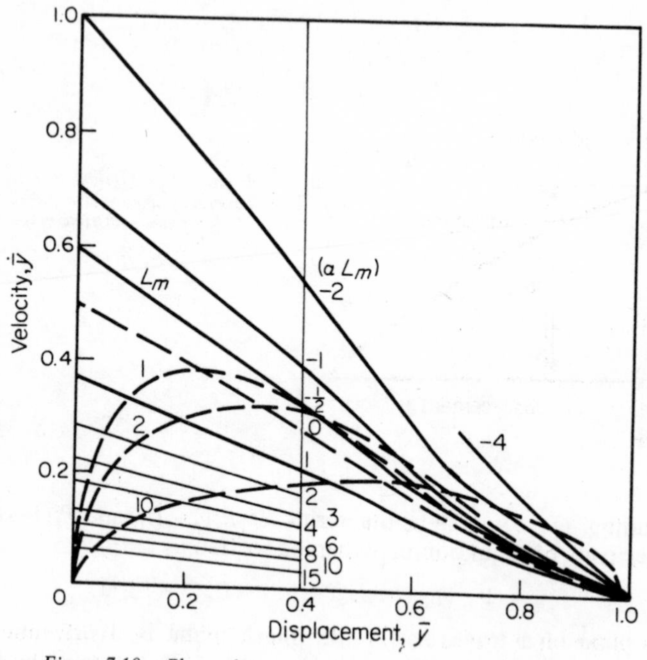

Figure 7.10. Phase plane step response when all loads are present

Curves (or isoclines) for a range of fixed values of αL_m are drawn as in *Figure 7.10* and values of L_m chosen. Let them be for example 1, 2 and 10. Each trajectory starts in the phase plane at $\dot{\bar{y}} = \bar{y} = 0$ and, since it has a finite acceleration at this point $d\dot{\bar{y}}/d\bar{y}$ must be infinite. The slopes with which the three trajectories cross the line $(L_m \alpha) = 15$, are 15, 7.5 and 1.5 respectively, and the entire response may, therefore, be constructed and transferred to the displacement-time plane by the method illustrated in *Figure 7.9*.

7.5 The sinusoidal response

A more popular method of testing and assessing the performance of servo systems is to measure their response to a sinusoidal input of constant amplitude over a wide range of frequencies. Whilst this method is of great value for linear systems it is virtually meaningless if considerable non-linear effects are present. Despite this, however, it is still in general use and it seems that it is unlikely to be replaced or discarded unless some 'standard', random form of input devised especially for testing purposes is generally accepted.

The transfer function is commonly used when considering sinusoidal responses and equation 7.10 may be written as

$$\frac{y}{x} = \frac{1}{(1+T\mathrm{D})} = \frac{1}{(1+j\omega T)} \qquad (7.11)$$

where the 'D' operator is replaced by the imaginary frequency term $j\omega$.

7.5.1 Gain, phase angle and Nyquist diagrams

Two forms of response curves are commonly used, one being the plot of the amplitude ratio, expressed in decibels. (*Note*: decibels are used to express the amplitude ratio and are given by $20 \log_{10}(y/x)$) against the non-dimensional form of the input frequency ωT and the other being an Argand diagram plot of the response in its real and imaginary parts with the variable again being the ωT parameter. Equation 7.11 is shown plotted in both these forms in *Figure 7.11*. The decibel plot is valuable since a simple and speedy approximation to it is given two straight lines, the first being horizontal and the second having a slope of 6db/octave. They meet at the 'break frequency' which is given by $\omega T = 1$ (where the error of the approximation is a maximum and equal to 3 db) and the phase angle or 'lag' of the system is shown to have a limiting value of $-\pi/2$.

The value of this method lies in the rapidity with which reasonably accurate approximations to the amplitude and phase angle curve of a multi-term transfer function can be drawn. For example, consider a transfer function of the form

$$\frac{y}{x} = \frac{(1+\mathrm{j}\omega T_1)}{(1+\mathrm{j}\omega T_2)(1+\mathrm{j}\omega T_3)}$$

where $T_1 > T_2 > T_3$.

The gain of the system in decibels is given by,

$$20\log_{10}(y/x) = 20\log_{10}(1+j\omega T_1) - 20\log_{10}(1+j\omega T_2) - 20\log_{10}(1+j\omega T_3)$$

and the gain or attenuation curve is then as shown in *Figure 7.12a*. At low frequencies $\omega T_1 < 1$ the response is constant or flat but for $\omega T_1 > 1$ the output amplitude or gain increases at the rate of 6 db/octave. When $\omega T_2 > 1$ the attentuation of the $(1+j\omega T_2)$ term

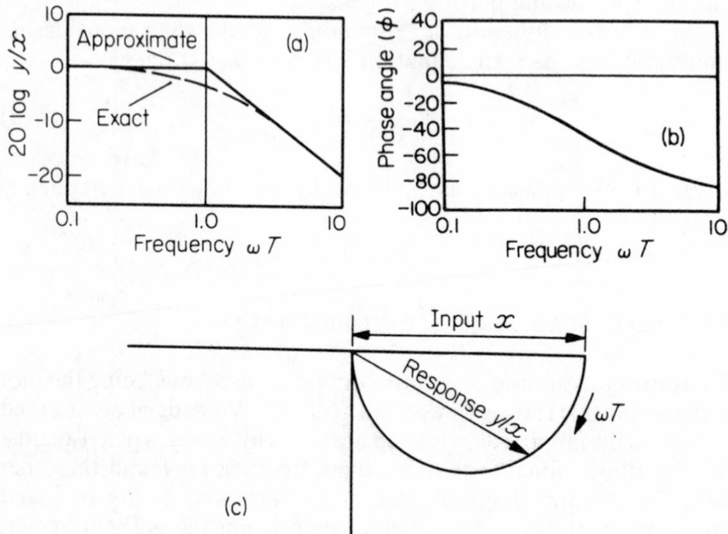

Figure 7.11. Linear, first order sinusoidal response. (a) Decibel amplitude—exact result; (b) Phase angle; (c) Argand diagram

begins to play a part in the response and cancels out any further gain due to the $(1+j\omega T_1)$ term. Increasing the frequency, ω, further, such that $\omega T_3 > 1$ then causes an overall attentuation of 6 db/octave. (*Note.* The words 'gain' and 'attenuation' are virtually synonomous, the former being preferred if the output is greater than the input, i.e., if amplification occurs or $y/x > 1$, whilst attenuation implies a reduction or $y/x < 1$.)

The phase angle curve can be drawn up in a similar manner and it is useful to note that the phase angles for a first order system having a transfer function of $1/(1+j\omega T)$ are approximately 5°, 45° and 85° at values of ωT of 0.1, 1 and 10 respectively as shown in *Figure 7.11b*.

The Argand plot of the response is called a Nyquist diagram and is useful since it shows both the amplitude and phase angle of the response using only one curve. Those for the functions $(1+j\omega T_1)$ and $1/(1+j\omega T_2)$ are shown in *Figure 7.13*. The former is a vertical straight line giving a gain which increases indefinitely with frequency

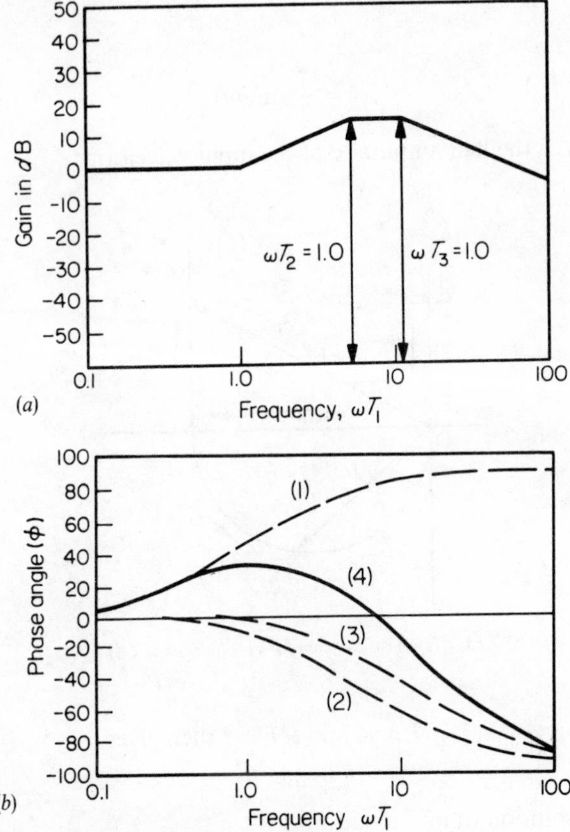

Figure 7.12. (a) Gain curves for system having three time constants; (b) Phase curves for system having three time constants

and a phase angle which approaches $\pi/2$ whilst the latter is a semi-circle showing a gain approaching zero and a phase angle approaching $-\pi/2$ as the frequency increases. Under some conditions it is also of value to consider the 'inverse' Nyquist diagram which, as its name implies, is an Argand diagram of x/y (see section 7.6.2).

7.5.2 The effect of load on the sinusoidal response

The sinusoidal response of the hydraulic relay was also derived in Reference 16 for the condition when the feedback loop was open and it was found convenient to compare the responses with that obtained for the no load condition,

i.e.

$$\dot{y} = \frac{X}{T}\sin(\omega t) \tag{7.12a}$$

where X is the half amplitude of the input waveform.

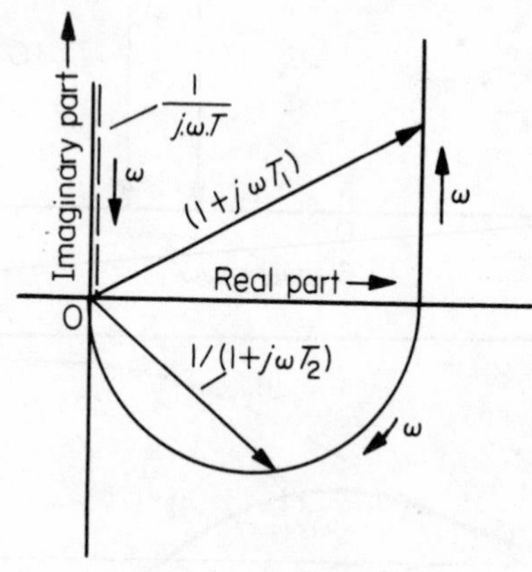

Figure 7.13. Nyquist diagram for $(1+j\omega T_1)$ *and* $1/(1+j\omega T_2)$

Writing $y/X = \bar{y}$, $\omega T = \bar{\omega}$ and $t/T = \bar{t}$ then gives

$$\dot{\bar{y}} = \sin(\bar{\omega}\bar{t})$$

and the solution to this equation is

$$\bar{y} = \frac{1}{\bar{\omega}}\{1-\cos(\bar{\omega}\bar{t})\} \tag{7.12b}$$

showing that the output has a sinusoidal form with a half amplitude, $\hat{\bar{y}}$, equal to $1/\bar{\omega}$ and that it lags the input by $\pi/2$. On the Nyquist diagram it is the vertical axis with frequency increasing towards the origin (broken line in *Figure 7.13*) and its transfer function may be written as

$$y/x = 1/\mathrm{j}\omega T$$

7.5.3 Constant load

As was shown for the step response, a system operating under constant load conditions is linear but the time constant T must be increased by the factor $1/\surd(1-L_f)$. If the non-dimensional, load parameter L_f equals unity the output of the system does not move.

7.5.4 Spring load

The response equation under these conditions is

$$\dot{\bar{y}} = \sin(\bar{\omega}\bar{t})\surd(1-L_s\bar{y}) \tag{7.13a}$$

and its solution may be written as

$$\cos(\bar{\omega}\bar{t}) = 1 - \frac{2\bar{\omega}}{L_s}\{1-\surd(1-L_s\bar{y})\} \tag{7.13b}$$

If the term $L_s\bar{y}$ is written as $\bar{\omega}\bar{y}L_s/\bar{\omega}$, then the above equation relates the response, expressed as $\bar{\omega}\bar{y}$, to a form of time, expressed as $\bar{\omega}\bar{t}$, in terms of a load and frequency parameter $\bar{\omega}/L_s$. Only one family of 'steady-state' response curves showing the variation of $\bar{\omega}\bar{y}$ with $\bar{\omega}\bar{t}$ need be drawn for various values of $\bar{\omega}/L_s$. Such a family is shown in *Figure 7.14*, where it can be seen that for high frequencies and/or small values of L_s, the response tends to that for the 'no-load' condition. For low frequencies or large values of L_s the response becomes attenuated, the wave form becoming more rectangular and the phase lag, ϕ, decreases from $\pi/2$ and approaches zero as $\bar{\omega}/L_s$ tends to zero. These characteristics are shown more clearly in *Figure 7.14b* and *c* where reciprocals of the load and frequency parameter are used when its value exceeds unity. Since the fundamental of a rectangular wave of unit height has a half amplitude $\hat{\bar{y}}$ of $4/\pi$ the curve shown in *Figure 7.14b* is corrected for this effect as $\bar{\omega}/L_s$ approaches zero. The values of the phase lag shown in *Figure 7.14c* are only approximate since they are the values of $\bar{\omega}\bar{t}$, expressed in degrees, at which the response first crosses the time axis. However, the values shown for $\bar{\omega}/L_s$ equal to zero and infinity have no error and it is unlikely that the maximum error for intermediate values is more than one or two degrees.

Figure 7.14d shows the variation of the half amplitude expressed as $\bar{\omega}\bar{y}$ as a function of the load and frequency parameter $\bar{\omega}/L_s$ in Nyquist diagram form.

Unlike the responses for other types of load, the spring-load response contains a transient term which gradually dies away. The 'steady-state' response can be obtained from the response equation 7.13a in the following manner. When the response is symmetrical about the time axis (i.e. 'steady state' conditions) the half amplitude of one half cycle will be equal to minus that of the following half cycle. The response equation may then be written in a more general form with the limits $\bar{y} = \bar{y}_o$ when $\bar{\omega}\bar{t}$ is equal to $\bar{\omega}\bar{t}_o$.

Integrating equation 7.13a then gives

$$\frac{2\bar{\omega}}{L_s}\{\sqrt{(1-\bar{\omega}\bar{y}L_s/\bar{\omega})}-\sqrt{(1-\bar{\omega}\bar{y}_o L_s/\bar{\omega})}\} = \cos\bar{\omega}\bar{t}-\cos\bar{\omega}\bar{t}_o$$

Since the response will have reached its half amplitude value of $\hat{\bar{y}}$ when $\bar{\omega}\bar{t}$ is equal to integral multiples of π the equation may be written as

$$\frac{2\bar{\omega}}{L_s}\{\sqrt{(1-\bar{\omega}\hat{\bar{y}}L_s/\bar{\omega})}-\sqrt{(1-\bar{\omega}\hat{\bar{y}}_o L_s/\bar{\omega})} = -2$$

With $\bar{\omega}\hat{\bar{y}} = -\bar{\omega}\hat{\bar{y}}_o$ and solving for $\bar{\omega}\bar{y}$ then gives

$$\bar{\omega}\hat{\bar{y}} = \sqrt{\{1-(L_s/2\bar{\omega})^2\}}$$

and this equation holds for $\bar{\omega}/L_s \geqslant 1/\sqrt{2}$. For values of $\bar{\omega}/L_s$ less than this, the value of $\omega\hat{\bar{y}}$ is equal to $\bar{\omega}/L_s$ since the output displacement is then limited by the spring.

7.5.5 Viscous load

When a viscous load is present the response equation becomes,

$$\dot{\bar{y}} = \sin(\bar{\omega}\bar{t})\sqrt{(1-L_v\bar{C}\dot{\bar{y}})} \qquad (7.14a)$$

After integrating and inserting the limits $\bar{y}=0$ when $\bar{t}=0$ the response is given by;

$$\bar{\omega}\bar{y} = \frac{R_L}{2}\left\{\frac{\sin(2\bar{\omega}\bar{t})}{2}-\bar{\omega}\bar{t}\right\}+\tfrac{1}{2}\{1-\cos(\bar{\omega}\bar{t})\}\sqrt{\{1+(R_L\sin(\bar{\omega}\bar{t}))^2\}}$$

$$+\frac{1+R_L^2}{2R_L}\left\{\sin^{-1}\left(\frac{R_L}{\sqrt{(1+R_L^2)}}\right)-\sin^{-1}\left(\frac{R_L\cos(\bar{\omega}\bar{t})}{\sqrt{(1+R_L^2)}}\right)\right\} \qquad (7.14b)$$

where $R_L = L_v/2$.

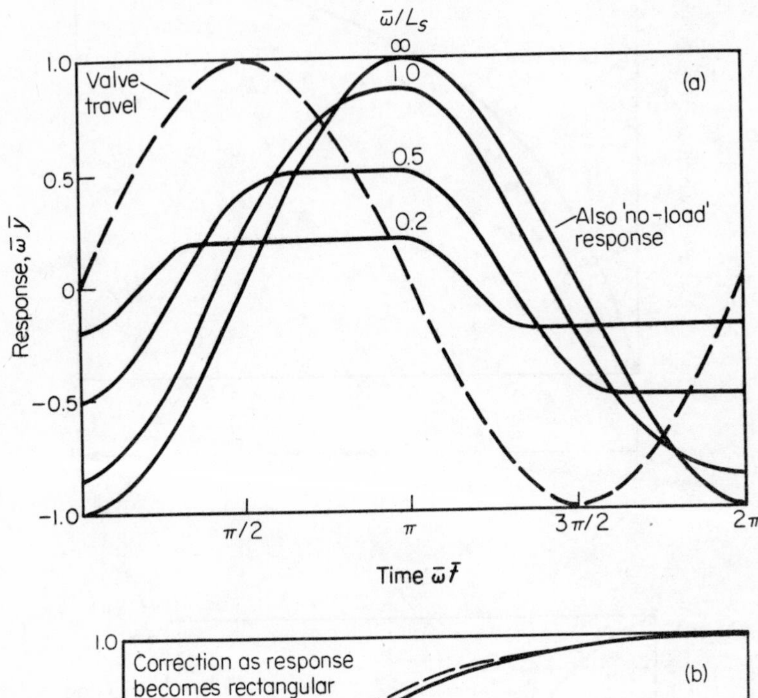

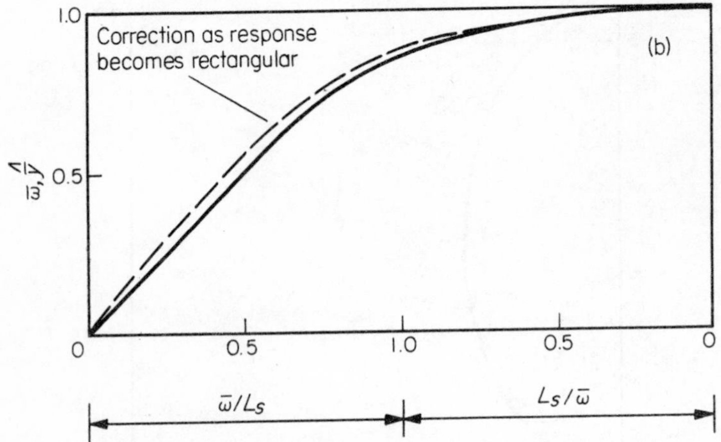

Figure 7.14. *Sinusoidal response with spring load*
(a) Waveforms
(b) Amplitude variation

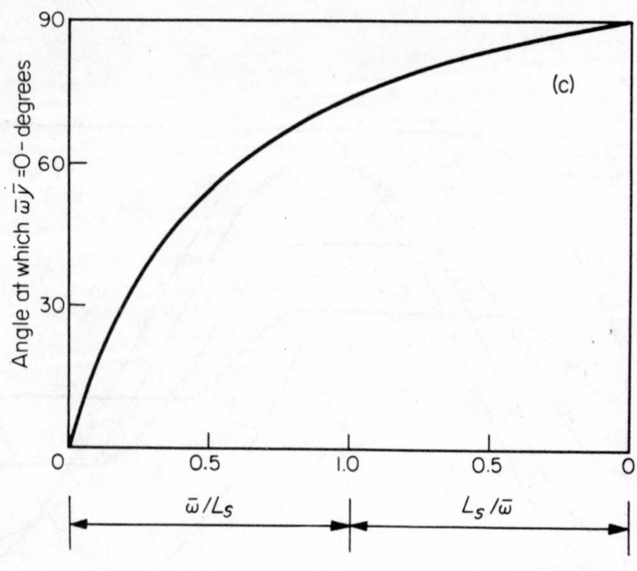

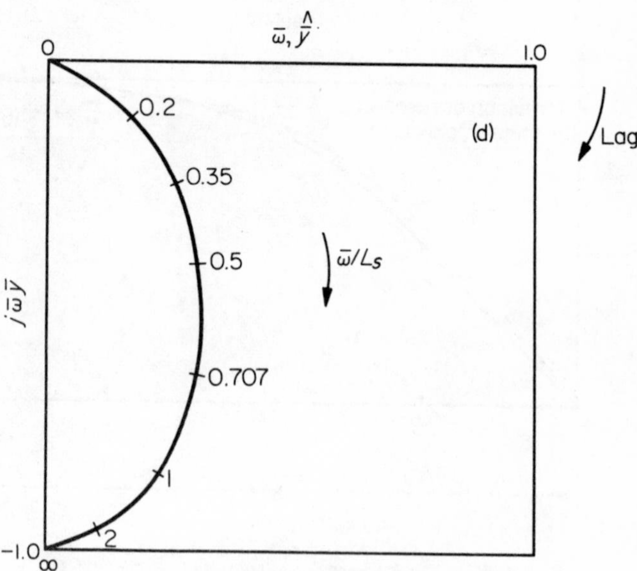

Figure 7.14 (continued)

(c) *Phase angle variation*
(d) *Nyquist diagram of response*

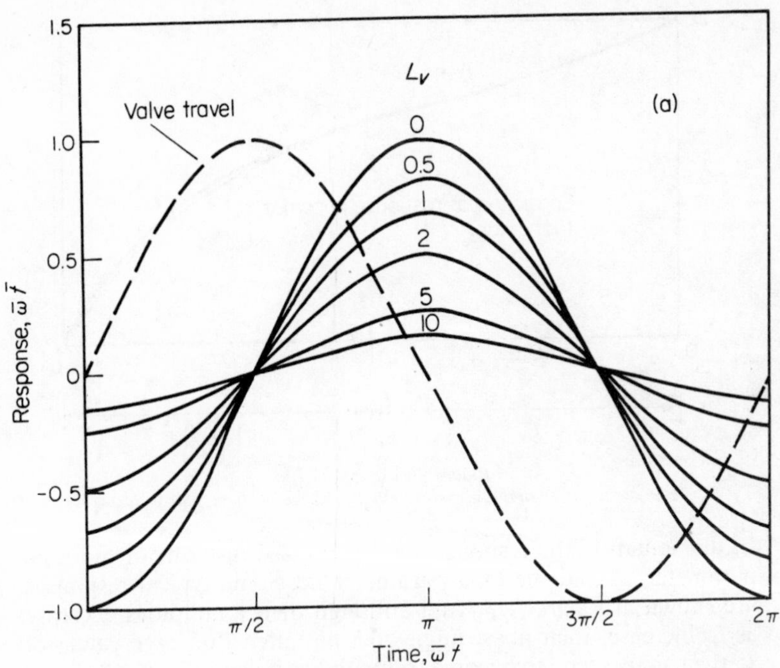

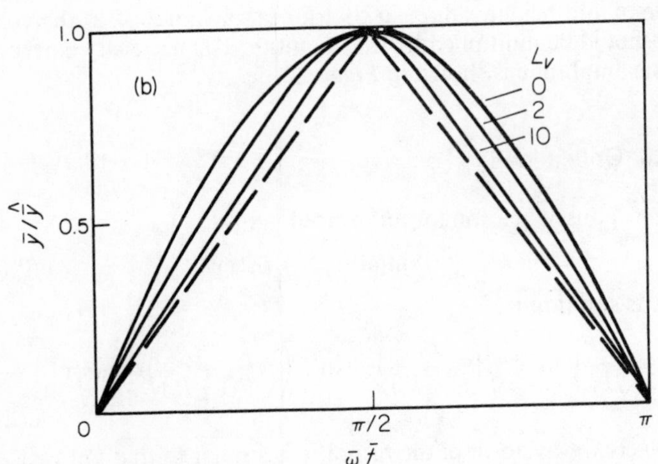

Figure 7.15. Sinusoidal response with viscous load

(a) Waveforms

(b) Variation of shape of response waveform

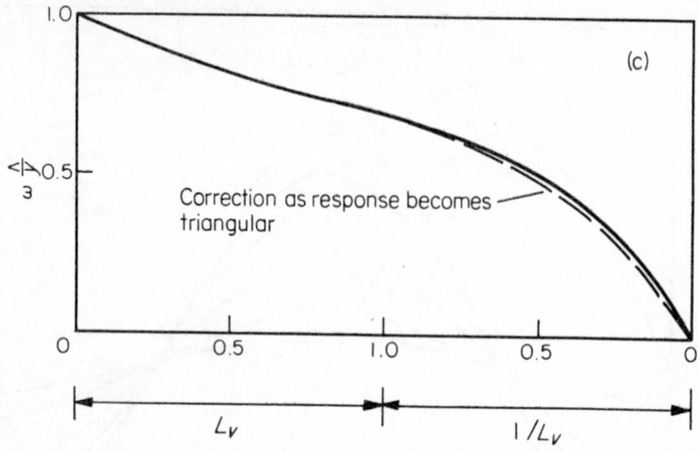

Figure 7.15 (continued)
(c) *Correction to amplitude ratio due to change of shape of waveform*

In this equation the response $\bar{\omega}\bar{y}$ is expressed in terms of the time parameter $\omega\bar{t}$ and the load parameter L_v. Some typical responses are shown in *Figure 7.15a* and although their amplitudes decrease as L_v increases their phase angle does not alter. For large values of L_v their shapes tend to become triangular as is shown in *Figure 7.15b* in which the response is shown as a fraction of its half amplitude, $\hat{\bar{y}}$. The half amplitude of the fundamental of a symmetrical triangular wave of unit height is $8/\pi^2$ so that for large values of L_v the values of $\bar{y}$ should be multiplied by this quantity. The necessary correction to the amplitude is shown in *Figure 7.15c*.

7.5.6 Orifice load

The response equation for an orifice load is

$$\dot{\bar{y}} = \sin(\bar{\omega}\bar{t})\sqrt{(1-L_r(\dot{\bar{y}})^2)} \qquad (7.15a)$$

and its solution is

$$\bar{\omega}\bar{y} = \frac{1}{\sqrt{L_r}}\left\{\sin^{-1}\sqrt{\left(\frac{L_r}{(1+L_r)}\right)} - \sin^{-1}\left(\sqrt{\left(\frac{L_r}{(1+L_r)}\right)}\cos(\bar{\omega}\bar{t})\right)\right\} \qquad (7.15b)$$

The actual waveform of the response is similar to that for a viscous load in that it tends to become triangular for large values of the load parameter but the change in shape takes place more slowly with increases in the load parameter.

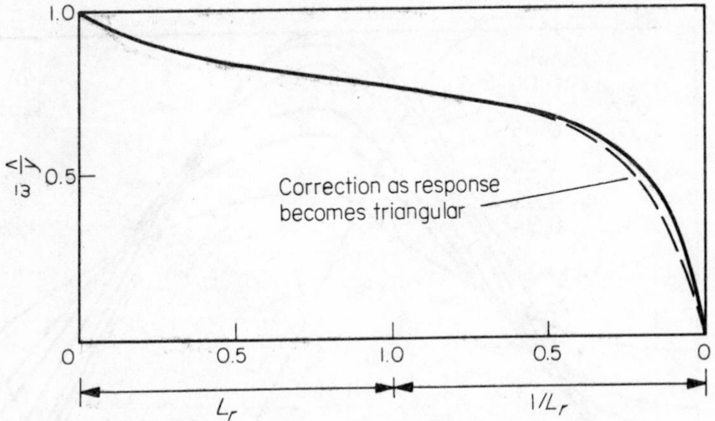

Figure 7.16. Correction to amplitude ratio due to change of shape of response waveform with orifice load

The necessary correction to allow for the change in waveform is shown in *Figure* 7.16 which gives the variation of the half amplitude of the response $\bar{\omega}\bar{y}$ with the load parameter L_r.

7.5.7 Inertial load

The response equation for an inertial load is

$$\dot{y} = \sin(\bar{\omega}\bar{t})\sqrt{(1 - L_m\ddot{y})} \qquad (7.16)$$

Curves of $\bar{\omega}\bar{y}$ as a function of $\bar{\omega}L_m$ are shown in *Figure 7.17a* and were derived in Reference 17 during the study of the effects of a combination of reaction forces and inertial loads on system stability. The variation of the response with this quantity is shown in Nyquist form in *Figure 7.17b* and it has a limiting phase lag of 122.4° as $\bar{\omega}L_m$ approaches infinity. It should be noted, however, that the errors produced by ignoring the harmonics are significant for values of $\bar{\omega}L_m$ exceeding unity.

7.5.8 Summary of results for sinusoidal responses

The presence of a spring load on the output of the system produces a non-linear response, and, for small values of the load frequency parameter, the output wave form becomes rectangular although, at the same time, the phase lag decreases.

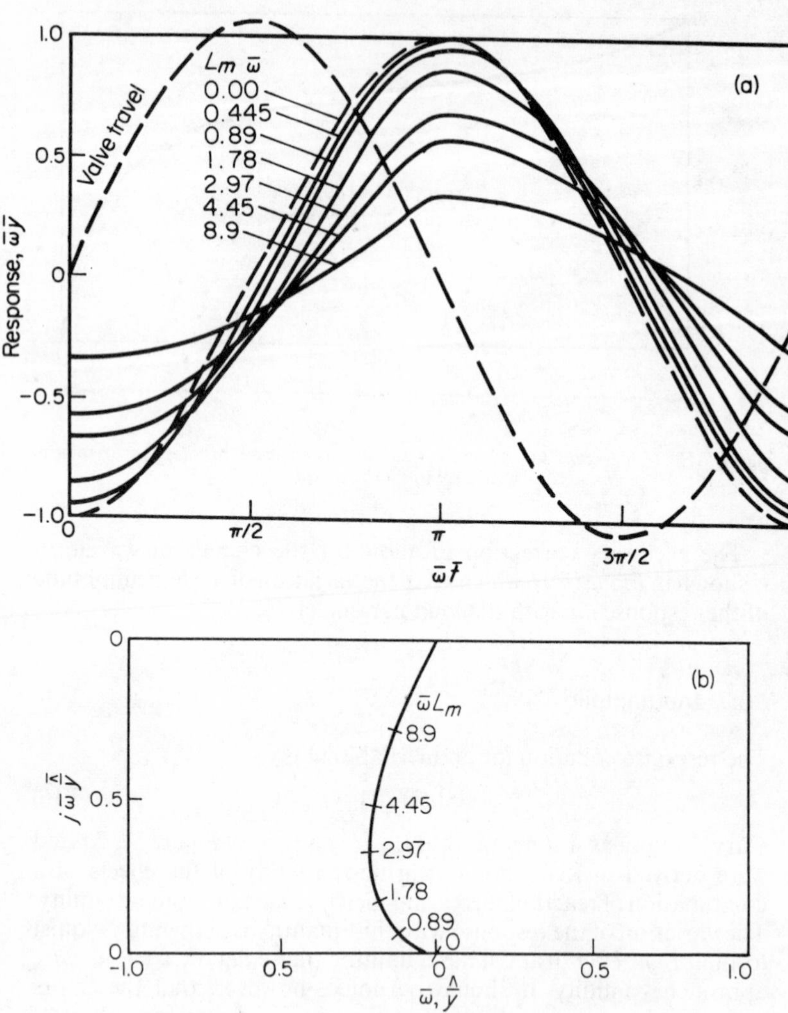

Figure 7.17. Sinusoidal response with inertial load
(a) Waveforms (b) Nyquist diagram of response with inertial load

The presence of either a viscous or an orifice load produces no change in the phase lag but attenuation occurs as the value of either the load or the frequency parameter increases. In both cases the velocity of the output tends to remain constant throughout each half cycle of the valve travel as would be expected from the nature of the loads.

As the inertial load on the system is increased more of the supply pressure is used to accelerate the load so that for a considerable portion of any half cycle of valve travel the acceleration is nearly constant. The displacement under such conditions is therefore parabolic as is shown for example by the curve for $\bar{\omega}L_m = 8.9$ in *Figure 7.17a*.

No sinusoidal results for combinations of loads have been considered but a graphical method, suggested by Huddlestone[18] could be used for such conditions.

7.6 Practical limitations of analyses

A considerable number of other factors influence the response of the hydraulic relay in practice and these include

Inertial loads and subsequent cavitation of the oil.
Oil compressibility.
Non-linear flow gain of valve (e.g., when circular ports are used).
Reaction forces acting on the valve spool.

7.6.1 Inertial loads and cavitation

The results derived in the preceding sections are valuable insofar as they indicate the variation of the response under various input conditions. When an inertial load is attached to the ram, however, it is essential to use caution when applying the results to predict actual responses. As an example of the effect of such loads it is of value to consider the effect of suddenly increasing and decreasing the valve opening of the system when the feedback linkage is removed.

If the valve is opened suddenly, the full supply pressure P_s acts on one side of the ram. In practice the exhaust or drain pressure P_e, of most systems is very small compared with P_s and may therefore be neglected. The initial acceleration of the ram is then given by $(P_s A_r)/M$, and, as the ram accelerates, the pressure drop, P_{v1}, across the supply control port and the pressure drop P_{v2}, across the exhaust control port both increase whilst P_r, the pressure drop across the ram, decreases. The pressures, P_1 and P_2, either side of the ram are given by $(P_s - P_{v1})$ and $(P_{v2} + P_e) \approx P_{v2}$ respectively. The variations of these pressures with time will be similar to that shown in *Figure 7.18*, the small difference between the steady state values of P_1 and P_2 being due to any friction forces acting on the ram and load.

Similarly a sudden reduction in valve opening causes the ram to decelerate, the necessary force being supplied by a transient drop in the pressure P_1 and a rise in the pressure P_2 as illustrated by the right-hand portion of *Figure 7.18*. The pressure drop across each port will exceed the approximate steady state value of $P_s/2$ until the speed of the ram equals the new value of the flow rate, Q (corresponding to the reduced valve opening) divided by the ram area.

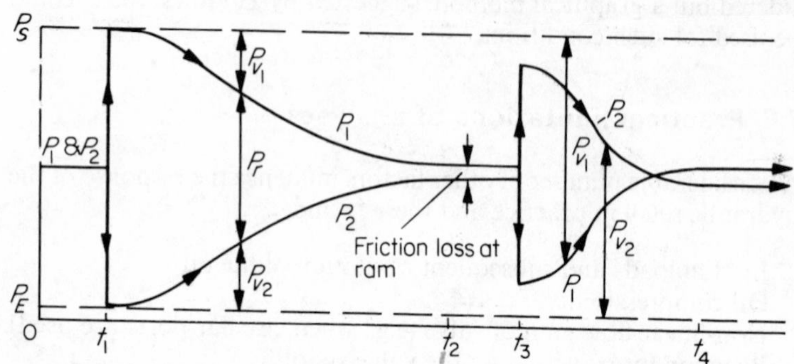

Figure 7.18. *Theoretical pressure variations for sudden spool movements*

P_s	*Supply pressure*
P_e	*Exhaust pressure* ≈ 0
P_1, P_2	*Pressures either side of ram.* P_1 *on supply side*
P_{v_1}, P_{v_2}	*Pressure drops across control ports.* P_{v_1} *on supply side*
P_r	*Pressure drop across ram*
t_1	*Valve suddenly opened*
t_2	*Output velocity reaches steady state*
t_3	*Valve opening suddenly reduced*
t_4	*Output velocity reaches steady state*

It should be noted that if the rate at which the valve closes is sufficiently high the pressure P_1 can easily drop to zero and attempt to become negative so that cavitation will occur on one side of the ram. If this does occur then, at the same time, the pressure P_2 will exceed P_s. References 14 and 15 contain a considerable amount of information on this subject and the latter contains an interesting study of the conditions accompanying the onset of cavitation in oil.

Using the method of isoclines the variation of the flow rate (ram velocity), the ram position and the pressures on either side of the ram were derived[13] for a sinusoidal input and gave the results shown in *Figure 7.19*. In parallel with this work Royle[14] conducted a series of experiments to study of the effects of cavitation and the associated behaviour of the system and two of the response traces he obtained are shown in *Figure 7.20*. The similarity of their form and those of the curves in *Figure 7.19b* and *Figure 7.19a* confirms that there is a reasonable degree of agreement between experimental and

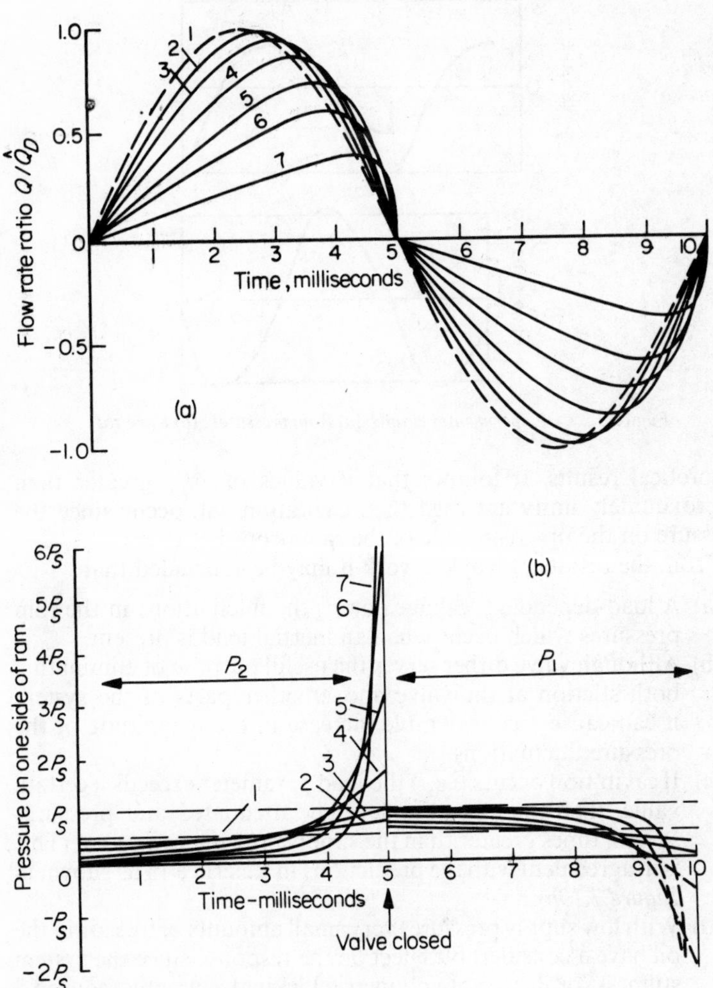

Figure 7.19. Characteristics of sinusoidal response with inertial load
(a) Flow rate (b) Ram pressures (frequency 100 Hz)

1. $\omega \cdot L_m = 0$
2. $\omega \cdot L_m = 0.445$
3. $\omega \cdot L_m = 0.89$
4. $\omega \cdot L_m = 1.78$
5. $\omega \cdot L_m = 2{\cdot}97$
6. $\omega \cdot L_m = 4.45$
7. $\omega \cdot L_m = 8.9$

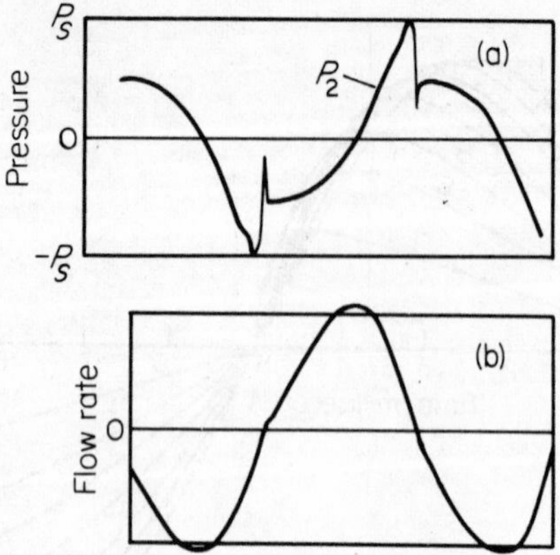

Figure 7.20. Experimental results. (a) Ram pressures; (b) Flow rate

theoretical results. It follows that if values of $\bar{\omega}L_m$ greater than approximately unity are used then cavitation will occur since the pressure on the upstream side of the ram drops below zero.

From the results of Royle's work it may be concluded that

(a) A load-dependent leakage damps the fluctuations in the ram pressures which occur when an inertial load is present.
(b) Although valve dither serves the useful purpose of eliminating both stiction at the valve and at other parts of the system it can cause a considerable increase in the magnitude of the pressure fluctuations.
(c) If cavitation occurs (i.e. if the load parameter exceeds a certain value) then the output is severely attenuated and pressures several times greater than the supply pressure can occur. (This is in agreement with the predictions in Refernce 13 as shown in *Figure 7.19b*).
(d) With low supply pressures very small amounts of free air in the oil have a considerable effect on the response since the system stiffness (or degree of volumetric locking) is greatly reduced.

During the experimental work of the study of spool valve instability[13] it was found that when the control ports of the four landed spool valve were 'short-circuited' by even a few cms of small diameter (6 mm) pipe the valve became unstable in a similar manner to that when it was driving a ram with an inertial load. This condition was examined in some detail but J. D. Stringer of the Department of Mechanical Engineering of Sheffield University has recently produced the following general analysis concerning an 'equivalent' load inertia of the oil in the pipework.

In the absence of all loads other than an inertial one on the ram and when the valve opening or supply pressure to the system suddenly changes the ram pressure drop, P_r, may be considered to consist of two parts, P_{r_1} and P_{r_2}. The pressure drop P_{r_1} is that required to accelerate the columns of oil in the pipes between the control ports of the valve and the inlet and outlet ports of the ram and P_{r_2} is the pressure drop required to accelerate the mass of the oil in the ram and the inertial load, M. In practice the mass of the oil in the ram is generally negligible compared with M and will be neglected although the necessary simple addition may be made if required.

For increasing valve openings and hence acceleration of the inertial load, M, on the ram

$$P_{r_2} = \frac{M}{A_r}\frac{\mathrm{d}\dot{y}}{\mathrm{d}t}$$

and, since the velocity of the oil in the pipes between the valve control ports and the ram ports is given by $A_r\,\dot{y}/a_1$ where a_1 is the cross sectional area of the pipes

$$P_{r_1} = \frac{M_1}{a_1}\frac{A_r}{a_1}\frac{\mathrm{d}\dot{y}}{\mathrm{d}t}$$

i.e.

$$\begin{aligned} P_r &= P_{r_1} + P_{r_2} = \{(M_1\,A_r)/a_1^2\}\,\mathrm{d}\dot{y}/\mathrm{d}t + (M/A_r)\,\mathrm{d}\dot{y}/\mathrm{d}t \\ &= \frac{M}{A_r}\frac{\mathrm{d}\dot{y}}{\mathrm{d}t}\left\{1 + \frac{M_1}{M}\frac{A_r^2}{a_1^2}\right\} \end{aligned}$$

In effect the inertial load, M, has been increased by the factor $(1 + M_1\,A_r^2/Ma_1^2)$. Since the square of an area ratio is involved this implies a diameter ratio raised to the fourth power but it must be noted that this is modified to some extent by the presence and position of the M_1 term, because this in turn is proportional to the pipe cross sectional area, a_1.

However, considering a small system in which for example only a total of 250 mm of pipework and valve block drillings are present having, say, a diameter of 3 mm and driving a load of 10 kg, this gives a mass ratio M_1/M of approximately 1/5000. If the ram diameter is 25 mm then the square of the area ratio, (A_r^2/a_1^2), is equal to 4800 so that the inertial load has been increased by a factor of approximately 2.0.

In the past, discrepancies between theoretical and experimental results concerning inertial load responses have often been attributed to inaccuracies in the estimation of the exact value of bulk modulus of the oil or working fluid. However, it would seem from the above that the effects of connecting pipes can be considerable and, in particular, when there is more than one change in the cross sectional area of such pipes (e.g., valve block drillings) the inertial load, M, of the system must be increased by the factor

$$\left\{1+\frac{M_1}{M}\frac{A_r^2}{a_1^2}+\frac{M_2}{M}\frac{A_r^2}{a_2^2}+\cdots\right\}$$

where a_1, a_2 etc., and M_1 and M_2 etc., are the cross sectional areas and masses of oil associated with the different diameters and lengths of connecting pipes. An alternative approach to a similar case concerning the stiffness or compliances of the oil in connecting pipes is derived by Walters[19].

7.6.2 Oil compressibility

Considering the system shown in *Figure 7.1* and the flow rate into the upstream side of the ram, then if the pressure on the upstream side of the ram is P_1, its value is given by

$$P_1 = P_s - P_v/2 \qquad (7.17)$$

since it will be assumed that the total valve pressure drop is P_v and that it is divided equally between the two ports.

The total flow rate Q_T into the high pressure side of the ram is given by

$$Q_T = A_r\,\dot{y} + \frac{V_1}{K'}\frac{\mathrm{d}P_1}{\mathrm{d}t} \qquad (7.18)$$

where K' is the bulk modulus of the oil (this is defined by $K' = V\,\mathrm{d}P/\mathrm{d}V$) and V_1 is the volume of oil at pressure P_1. The compressibility term is required to allow for the small quantity of oil required to raise the pressure by $\mathrm{d}P_1$.

Other equations required have been previously derived (see section 7.12) and these are

$$z = (x-y) \qquad (7.2)$$

$$Q = kz\sqrt{P_v} \qquad (7.3)$$

$$P_r = P_s - P_v = M\ddot{y}/A_r \qquad (7.1 \text{ and } 7.4)$$

It is convenient to write

$$\left.\frac{\partial Q}{\partial z}\right|_{P_v \text{ const}} = C_z \tag{7.19}$$

and

$$\left.\frac{\partial Q}{\partial P_v}\right|_{z \text{ const}} = C_p \tag{7.20}$$

(*Note*. The value of C_p defined by this equation is half the magnitude of that used by Harpur which is equal to $-\partial Q/\partial P$, see pages 177 and 178)

and from equation 7.17 $\quad dP_1 = -dP_v/2 \quad$ (7.21)

No analytical solution for y as a function of x is known but Harpur[20], showed that by using the method of small perturbations the stability of a small variation about a steady-state condition may be considered as follows. Let each quantity increase by a very small amount such that they may be replaced by their increased valves as follows,

z by $z+z_1$

x by $x+x_1$

y by $y+y_1$

Q by $Q+q = Q+C_p p_v+C_z z_1$

P_v by P_v+p_v

Rewriting the equations in the manner

$$z+z_1 = (x+x_1-y-y_1)$$

and subtracting the original equations

$$(\text{i.e. } z = x-y)$$

then gives

$$z_1 = x_1-y_1$$

$$q = C_p p_v+C_z z_1$$

$$q = A_r \dot{y}_1+\frac{V_1}{K'}\frac{dp_1}{dt} = A_r \dot{y}_1-\frac{V_1}{2K'}\frac{dp_v}{dt}$$

$$p_v = -\frac{M}{A_r}\ddot{y}_1$$

or

$$dp_v/dt = -\frac{M}{A_r}\dddot{y}_1$$

Equating the two expressions for q above and substituting for z_1. p_v and dp_v/dt then gives

$$-C_p\frac{M}{A_r}\ddot{y}_1+C_z(x_1-y_1) = A_r \dot{y}_1+\frac{V_1}{2K'}\frac{M}{A_r}\dddot{y}_1$$

Considering the ram in its central position (see Reference 19 if this is not so) such that $V_1 = V_T/2$ where V_T is the total ram volume finally gives

$$(V_T M/2K'A_r)\dddot{y}_1 + C_p(M/A_r)\ddot{y}_1 + 2A_r\dot{y}_1 + 2C_z y_1 = 2C_z x_1 \quad (7.22)$$

or the transfer function y_1/x_1 is given by

$$y_1/x_1 = \frac{C_z}{(V_T M/4K'A_r)D^3 + C_p(M/A_r)D^2 + A_r D + C_z} \quad (7.23)$$

The condition for stability is that the product of the coefficients of the D^2 and D terms in the denominator is greater than that of the D^3 and the constant term, i.e.

$$(C_p M/A_r)A_r > (V_T M/4K'A_r)C_z$$

or

$$(C_p/C_z) > (V_T/4K'A_r) \quad (7.24)$$

It is important to note that this condition will not be satisfied for small valve travels (small values of z) since C_p is directly proportional to z as is shown by equation 6.4 in Chapter 6. In general, it is desirable that when the system has a sinusoidal input, say $x_1 \sin(\omega t)$, the maximum value of the output ratio y/x should not exceed 1.4 but to simplify the working a value of unity will be considered. Under these conditions equation 7.23 may be written as

$$1 = \frac{C_z}{\{-(V_T M/4K'A_r)(j\omega^3) - (C_p M/A_r)\omega^2 + A_r j\omega + C_z\}}$$

This gives

$$1 = \frac{C_z}{\sqrt{(\{C_z - (C_p M/A_r)\omega^2\}^2 + \omega^2\{A_r - (V_T M\omega^2/4K'A^r)\}^2)}}$$

or

$$1 = \frac{1}{\sqrt{\{(1-b\omega^2)^2 + \omega^2(c-a\omega^2)^2\}}} \quad (7.25)$$

where $a = V_T M/4K'A_r C_z$; $b = MC_p/A_r C_z$ and $c = A_r/C_z$.

At the maximum amplitude point (ω varying) the differential coefficient of this expression must be zero, giving

$$3a^2\omega^4 + (2b^2 - 4ac)\omega^2 + (c^2 - 2b) = 0 \quad (7.26)$$

and for equation 7.25 to hold

$$a^2\omega^4 + (b^2 - 2ac)\omega^2 + (c^2 - 2b) = 0 \quad (7.27)$$

Solving for ω^2 in both these equations and equating the values gives

$$\pm 2\sqrt{(1-3\psi')} = -1 \pm 3\sqrt{(1-4\psi')}$$

where $\psi' = a^2(c^2 - b^2)/(b^2 - 2ac)^2$

and solving for ψ' gives $\psi' = 0$ or 1/4 so that for a maximum amplitude ratio of unity the following two equations are obtained

$$\frac{a^2(c^2-2b)}{(b^2-2ac)^2} = 0 \tag{7.28}$$

and

$$\frac{a^2(c^2-b^2)}{(b^2-2ac)^2} = \tfrac{1}{4} \tag{7.29}$$

Combining equations 7.28 and 7.29 gives the frequency at which the amplitude ratio is unity as follows

$$\omega = \sqrt{\left(\frac{2ac-b^2}{2a^2}\right)} \tag{7.30}$$

and since from equation 7.28 $a^2 = 0$ or $c^2 = 2b$ this means that either the frequency is infinite, which is of no interest, or

$$C_p C_z = A_r^3/2M \tag{7.31a}$$

Figure 7.21a shows the form of this hyperbolic relationship between C_p and C_z. The other equation, 7.29, gives

$$b^3 - 4abc + 8a^2 = 0 \tag{7.31b}$$

and converting this into a relation between A_r, M, K and V_T and the parameters C_p and C_z gives the elliptical form of curve also shown in *Figure 7.21*. All points inside this curve will give amplitude ratios less than unity and the slope of this curve at the origin is given by

$$C_p/C_z = V_T/2K'A_r \tag{7.32a}$$

which is exactly twice the value necessary for stability given by equation 7.24, and which is also shown in *Figure 7.21a*. Harpur also derived the responses for systems having values of C_p and C_z represented by the points A, B, C and D in *Figure 7.21a* and these are shown in *Figure 7.21b*. All these systems give satisfactory results, but only system A has an amplitude ratio which never exceeds unity and this would generally be regarded as an 'over-cautious' approach.

If the amplitude ratio is fixed at 1.4 or $\sqrt{2}$ then equation 7.32a becomes

$$C_p/C_z = 7V_T/16K'A_r \tag{7.32b}$$

and this stability limit is generally considered to be of more practical value.

For given values of spool travel, curves can be drawn relating C_p to C_z and lines of constant values of pressure drop across the ram, P_r are vertical so that the entire region of a system's operating condition can be drawn as shown in *Figure 7.22*, where values of an actual system's various parameters have been used.

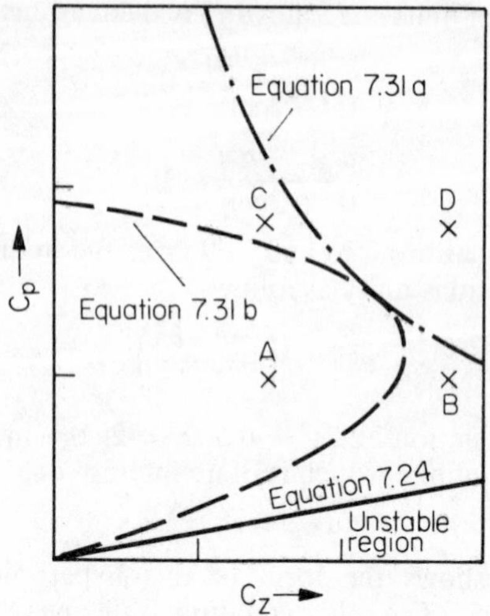

Figure 7.21(a). C_p/C_z diagram for system allowing for oil compressibility

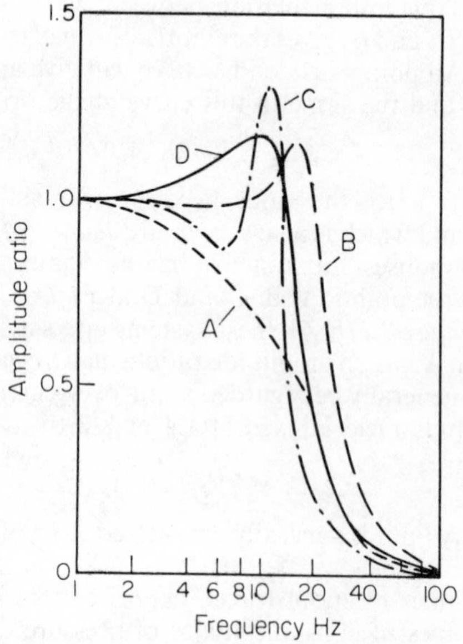

Figure 7.21(b). Frequency response curves for points A, B, C and D

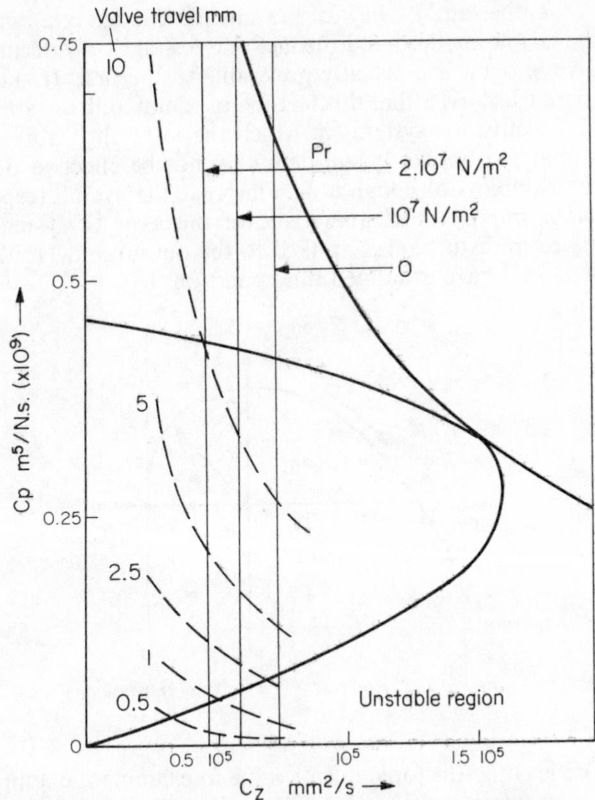

Figure 7.22. Variation of C_p with C_z and lines of constant pressure and valve travel

It can be seen that, below a certain value of spool travel, the system under consideration is *bound to be unstable* since C_p falls to zero when the error and ram pressure drop approach zero. However, although this shows that any valve having zero or positive lap is theoretically prone to instability, the instability may not necessarily occur in practice and several factors (such as ram friction) which have been neglected may reduce the predicted band of instability to a reasonably small or even negligible area. A similar conclusion has been drawn by Faisandier[21] and by Friedrich[22] who have also extended this type of analysis to examine the effects of component and structural elasticity.

Harpur shows that two methods of increasing the stability of such systems may be used, the first is to introduce a small amount of underlap and the second is to insert a small, orifice-type, leakage path across the ram. In view of the smaller leakage rate associated with the second method and the fact that leakage only occurs when the valve is open it is usually preferable to the first. It should be emphasised, however, that this method of eliminating instability will only be effective for systems in which the instability is due to the effects analysed above. It will definitely not be effective on other types of instability although it may improve the overall response.

Another important characteristic of this type of system is its impedance to disturbances applied to the output and Houlobek[23] and Watson[24] have examined this condition.

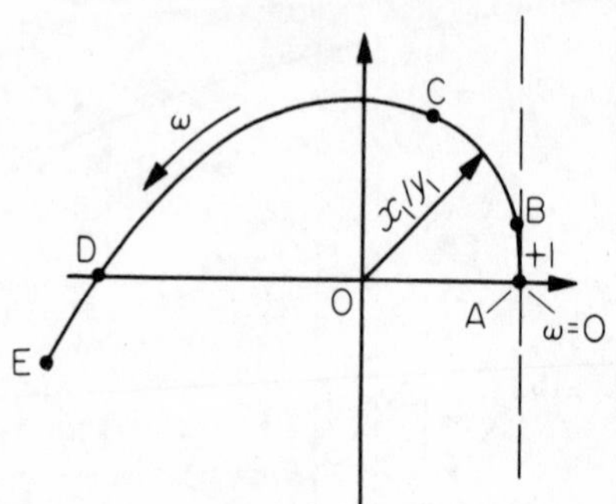

Figure 7.23. Inverse Nyquist plot of response

Before leaving this topic it is of value to examine the approaches made by Royle[12] and Faisandier[21] in the latter of which equation 7.23, is inverted and the right hand side is divided throughout by C_z. This gives

$$\frac{x_1}{y_1} = 1 + \frac{A_r}{C_z} D + \left(\frac{C_p M}{C_z A_r}\right) D^2 + \left(\frac{V_T M}{4K' A_r C_z}\right) D^3 \qquad (7.23a)$$

For sinusoidal operation, $D = j\omega$, and the above equation may then be written as

$$\frac{x_1}{y_1} = 1 - \frac{C_p M}{C_z A_r}\omega^2 + j\omega \frac{A_r}{C_z}\left\{1 - \frac{V_T M}{4K' A_r^2}\omega^2\right\} \qquad (7.23a)$$

The *natural frequency*, ω_n, of the mass and ram part of the system may be written as

$$\frac{1}{\omega_n^2} = \frac{V_T M}{4K' A_r^2}$$

so that the frequency response equation then becomes

$$\frac{x_1}{y_1} = 1 - \frac{C_p M}{C_z A_r}\omega^2 + j\omega\frac{A_r}{C_z}\left\{1 - \frac{\omega^2}{\omega_n^2}\right\} \tag{7.23b}$$

and its inverse Nyquist diagram is shown in *Figure 7.23*. At very low frequencies ($\omega \to 0$), the inverse amplitude ratio approaches unity (point A) and initially as ω increases the term $j\omega A_r/C_z$ gives the vertical, AB, portion of the curve. For higher frequencies the second term of the right hand side of equation 7.23 produces a negative but real component (see portion BC of curve) and eventually the $j\omega^3$ term produces the CDE portion.

Stability is assured if the point D lies to the left of the origin as shown and at this point, from equation 7.23a, $\omega = \omega_n$ thereby reducing the imaginary part of the equation to zero. The value of x_1/y_1 is then equal to $1-(C_p M\omega_n^2/C_z A_r)$ and for this to be negative

$$\frac{C_p M}{C_z A_r}\omega_n^2 > 1 \quad \text{or} \quad \frac{C_p}{C_z} > \frac{A_r}{M\omega_n^2} \tag{7.23c}$$

Inserting the value of $1/\omega_n^2$ given above then yields

$$\frac{C_p}{C_z} > \frac{V_T}{4K'A_r} \tag{7.23d}$$

which is in agreement with equation 7.24 derived above.

From chapter 6 the values of C_z and C_p are given by equations 6.3 and 6.4 and are

$$C_z = C_d w\sqrt{(P_v/\rho)} \quad \text{and} \quad C_p = C_d wz/2\sqrt{(P_v\rho)}$$

so that the ratio C_p/C_z is given by

$$\frac{C_p}{C_z} = \frac{z}{2P_v} \tag{7.23e}$$

Now the ratio V_T/A_r is approximately equal to the total stroke of the ram and taking for example this to be 125 mm, the effective bulk modulus of the oil K' to be 13 000 bars and the total valve pressure drop, P_v, to be 70 bars then gives from equations 7.23c and 7.23d.

$$\frac{z}{2P_v} > \frac{V_T}{4K'A_r} \quad \text{or} \quad z > \frac{2P_v V_T}{4K'A_r}$$

i.e. $z > 0.3$ mm.

This implies instability for valve travels of less than 0.3 mm, and to avoid this by the use of valve underlap would require the use of a valve underlap of this magnitude. In most systems such an underlap would be unacceptable.

The following operating conditions are also of interest in view of their influence on the stability criterion:

(a) The presence of a viscous leakage path across the ram the flow rate of which is given by $K_L P_r$. This modifies the small perturbation analysis such that the stability criterion of equation 7.23c becomes,

$$\frac{C_p + K_L}{C_z} > \frac{V_T}{4K' A_r} \tag{7.23f}$$

which can be satisfied even when z and hence C_p approaches or is equal to zero.

(b) The presence of a spring between the ram and the inertial load with the feedback being taken from the position of the ram and not that of the load[18,21]. Under this condition the original stability criterion, given by equation 7.23c, still holds but it is important to ensure that the inertial load M does not resonate in conjunction with the spring.

(c) If the ram housing is supported on an elastic member of stiffness s where s is large but not infinite and if both the resulting fourth and fifth order terms in ω are neglected then the small perturbation response equation is stable if

$$\frac{C_p}{C_z} > \frac{V_T}{4K' A_r} + \frac{A_r}{s} \tag{7.23g}$$

which indicates a condition which is more difficult to fulfill than that indicated by equation 7.23c.

(d) If both the ram and valve are housed in one body and this is supported on an elastic member then for the same conditions as given in (c) above the system is stable if

$$\frac{C_p}{C_z} > \frac{V_T}{4K' A_r} - \frac{A_r}{s}$$

which is a condition more easily satisfied than that given by 7.23c.

(e) The presence of both an inertial load M and a viscous load μ' such that the force required at the ram is given by $P_r A_r = M\ddot{y} + \mu' \dot{y}$ produces the following small perturbation response equation

$$\frac{x_1}{y_1} = 1 - \left\{\frac{C_p M}{C_z A_r} + \frac{V_T \mu'}{4K' A_r C_z}\right\}\omega^2 + \mathrm{j}\omega \frac{A_r}{C_z}\left\{\left(1 + \frac{C_p \mu'}{A_r^2}\right) - \frac{\omega^2}{\omega_n^2}\right\}$$

and the condition for stability is

$$\left(\frac{C_p M}{C_z A_r}+\frac{V_T \mu'}{4K' A_r^2 C_z}\right)\left(1+\frac{C_p \mu'}{A_r^2}\right)>\frac{1}{\omega_n^2}$$

As z approaches zero the term $C_p \mu'/A_r^2$ becomes small compared with unity and this condition may then be written as

$$\frac{C_p}{C_z}>\frac{V_T}{4K'}\left\{\frac{1}{A_r}-\frac{\mu'}{C_z M}\right\} \tag{7.23h}$$

which again is a condition more readily satisfied than that called for by equation 7.23c.

From the various conditions considered it would appear that valve underlap may be discarded for the reason already stated, the use of (a) is of value, the use of (b) has no effect and the use of (c) makes stability more difficult to achieve, although with the prescribed large value of s and hence a small value of A_r/s the change may be marginal. With condition (d) the change whilst beneficial may, for the same reasoning applied to condition (c), provide only marginal benefit whereas condition (e) suggests considerable improvement may be obtained if the value of μ' is adequate.

As a result conditions (a) and (e) hold out distinct possibilities of improving a system's stability and it is a combination of these which Royle[12] studied both theoretically and experimentally in connection with the response of a machine tool table at low rates of feed. His aim was to reduce the 'stick-slip' component of the motion of the table and it was found that by using a viscous leak across the ram and heavy load damping at low speeds a mass of 100 kg could be positioned to within 2.10^{-3} mm, of the desired position whereas a mean error in position of 5.10^{-2} mm existed previously.

7.6.3 Non-linear valve characteristics

The 'flow rate/valve travel' characteristic shown in *Figure 6.7a* of the previous chapter was drawn for a valve having rectangular ports and a constant discharge coefficient. It will be recalled that the variations in discharge coefficient, particularly at small spool travels, were discussed together with a method of reducing their effect on the system by using valves with large travels. If circular ports are used, however, the shape of the port area/spool travel characteristic is as shown in *Figure 7.24* and the resulting variation in the flow gain coefficient (for circular ports this will be denoted by C_z') must be considered since it approaches zero as the spool travel decreases. The effects produced on the response by asymmetrical lap have been studied in Reference 14 of the previous chapter.

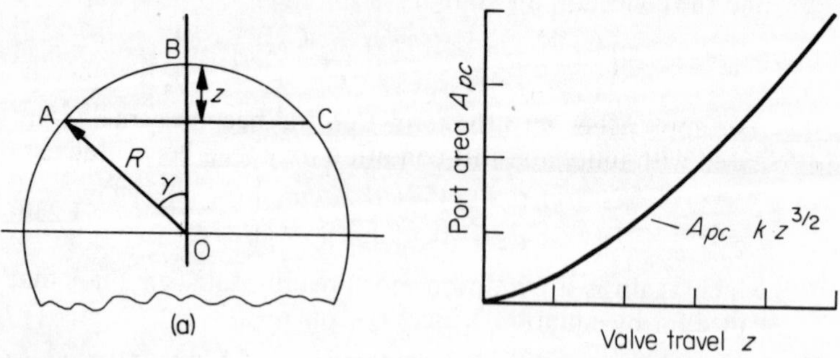

Figure 7.24. *Port area of a valve with circular ports. (a) Port configuration, port area ABC; (b) Area/travel characteristic*

7.6.4 The effect of circular ports on system response

The relative ease with which circular ports may be inserted in the sleeve of a valve compared with rectangular ones makes them extremely popular from the manufacturing point of view, since their use generally reduces the production costs.

A detailed study of the effect of spool and liner clearance on the effective port area of a valve having circular ports is given in Reference 25 for two particular values of the clearance ratio. It was found that the effective area of the port A'_{pc} is given by

$$A'_{pc} = \frac{1.5}{CR} A_{pc} \tag{7.33}$$

where A_{pc} is the port area when there is no clearance and CR is the ratio of spool travel/radial clearance in the region of the port. From the results obtained it appears that equation 7.16 holds for values of the clearance ratio up to approximately 0.5.

If these clearance effects, which can generally be made very small for most of the spool travel, are neglected then from *Figure 7.34* the port area A_{pc} is given by

$$A_{pc} = R^2(\gamma - \tfrac{1}{2}\sin 2\gamma) \tag{7.34}$$

and eliminating γ gives

$$A_{pc} = \pi R^2 f(z/R)$$

where $f(z/R) = \{\cos^{-1}(1-z/R) - (1-z/R)[z/R(2-z/R)^{1/2}]\}/\pi$.

The flow rate Q is given by

$$Q = n' C_d A_p \sqrt{(P_s/\rho)}$$

where n' is the number of control ports and the output velocity dy/dt is then given by

$$\frac{\mathrm{d}y}{\mathrm{d}t} = \frac{Q}{A_r} = \frac{n' C_d \pi R^2 f(z/R)}{A_r} \sqrt{\left(\frac{P_s}{\rho}\right)} \tag{7.35a}$$

which may be written as

$$\frac{\mathrm{d}\bar{y}_1}{\mathrm{d}\bar{t}} = f(\bar{z}_1)$$

where $\bar{z}_1 = (\bar{x}_1 - \bar{y}_1)$, $\bar{x}_1 = x/R$, $\bar{y}_1 = y/R$, $\bar{z}_1 = z/R$ and $t/T_c = \bar{t}_c$ with the system time constant, T_c, given by

$$T_c = \frac{A_r}{n' C_d \pi R} \sqrt{\left(\frac{\rho}{P_s}\right)}$$

7.6.5 The step response with circular ports

If a step input of magnitude X is applied to the system then the response may be described by the integral of equation 7.35a which may be written as

$$\bar{t}_c = \frac{1}{\pi} \int_{\eta_0}^{\eta} f(\eta')\,\mathrm{d}\eta'$$

where $f(\eta') = \sin\eta'/(\eta' - \sin\eta'\cos\eta')$, $\eta' = \cos^{-1}(1 - \bar{z}_1)$ and η'_0 is the initial value of η' at $\bar{t}_c = 0$.

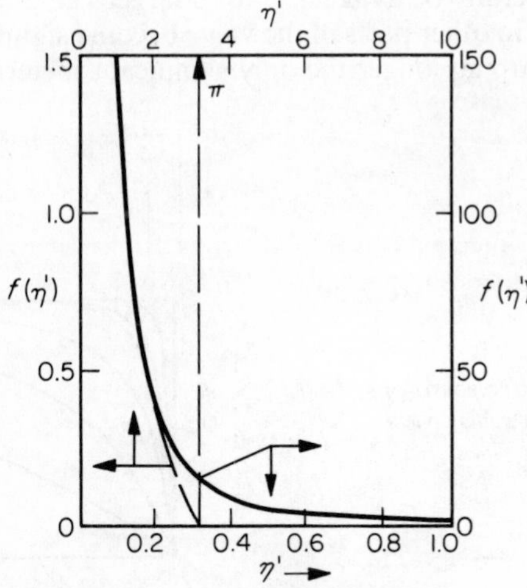

Figure 7.25. Variation of $f(\eta')$ with η'

No analytical solution to this integral is known and it is therefore necessary to plot $f(\eta')/\eta'$ as shown in *Figure 7.25* with a view to graphical integration. For values of η' between 2.2 and (n') graphical integration must be used but if $\eta' < 2.2$ then it can be shown that

$$f(\eta') \approx 3/2(\eta')^2$$

However, other limitations complicate this method of approach in the range where this approximation is valid and it is simpler to return to equation 7.34 when $\eta' < 2.2$. Assuming $z/R < 1$ and using series for the $\cos^{-1}(1-z/R)^{1/2}$ terms gives

$$A_p = \frac{4\sqrt{2}}{3} R^2 (\bar{z})^{3/2}$$

if powers of $\bar{z}$ greater than 2 are neglected.

The response equation may then be written as

$$\frac{d\bar{y}_1}{d\bar{t}'_c} = (\bar{X} - \bar{y}_1)^{3/2} \tag{7.35b}$$

where $\bar{t}'_c = t/T'_c$, $T'_c = 3A_r\sqrt{\rho}/n'C_d 4\sqrt{2}R\sqrt{P_s}$, $\bar{y}_1 = y/R$ and $\bar{X} = \text{Input}/R$.

Equation 7.35b may be integrated with the limits $\bar{y} = \bar{t}'_c = 0$ giving

$$\frac{y}{x} = 1 - \frac{4}{(\bar{t}_c\sqrt{\bar{X}} + 2)^2} \tag{7.36}$$

and this gives accurate results (within 5%) for $x/R \leqslant 0.2$.

It should be noted that in practice values of x/R greater than 0.2 should generally be avoided, since if larger values are used then the flow losses in other parts of the valve become significant and the control ports are no longer the only significant metering devices in

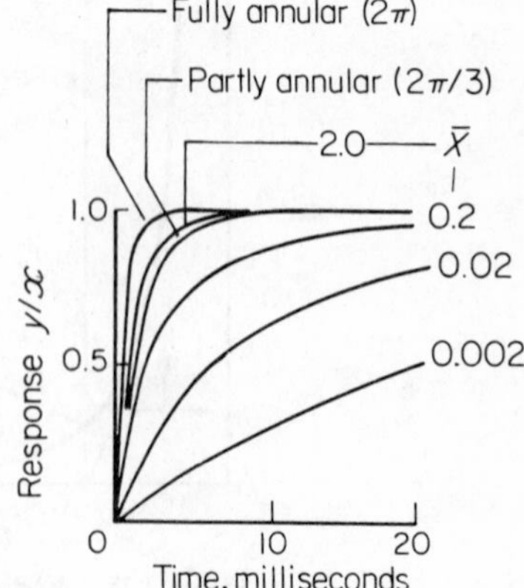

Figure 7.26. Step responses of system with valve having circular ports

the hydraulic circuit. As a result this simple approximation is generally quite acceptable.

Examples of the response are shown in *Figure 7.26* and may be compared with those of systems having valves with rectangular ports. For the condition $x/R > 0.2$ a reasonably accurate result is obtained if a linear approximation to the port area/valve travel characteristic is taken and a response curve for $x/R = 2.0$ is also shown in *Figure 7.26* where the $4\sqrt{2}\,nR/3$ term in the time constant equation has been replaced by the slope of the linear approximation to the port area curve, $\pi nR/2$.

7.6.6 The sinusoidal response with circular ports

Several methods of solving the sinusoidal response equation have been examined in Reference 25 and it was found that very satisfactory results can be obtained by the 'describing function' method which has been developed by Grief[26]. This method is based on the

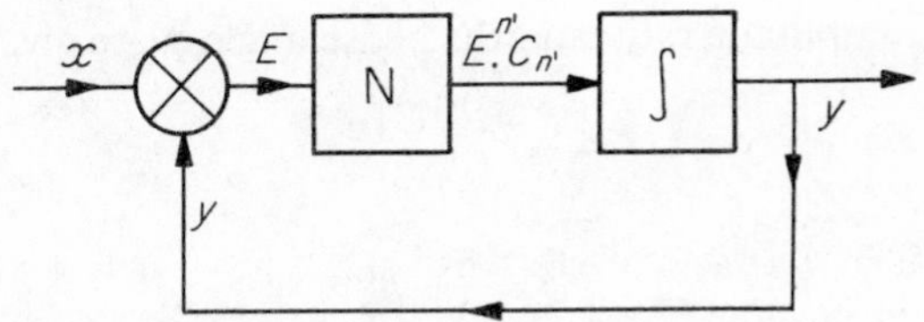

Figure 7.27. System analysed by describing function method

assumptions that the signal fed into the non-linear element (in this case the valve having a non-linear, port area/spool travel characteristic) is sinusoidal, that the output from this element may be satisfactorily represented by the fundamental of the Fourier series for the output and that the system, on open loop, acts as a low pass filter, thereby reducing the harmonics. This latter condition is very important and is applicable in this case since the ram is an integrator and as a result the nth harmonic will be attenuated by the factor $1/n$.

The system shown in *Figure 7.27* consists of an error measuring device, an nth power law, non-linear element N and an integrator. The error of the system ε will be given by

$$\varepsilon = E\sin(\omega t) \tag{7.37}$$

and the input x by

$$x = A'\sin(\omega t + \beta') \tag{7.38}$$

Grief has shown that the fundamental of the output from the non-linear element of the system may be written as $E^{n'}C_{n'}\sin(\omega t)$ where $C_{n'}$ is a coefficient the magnitude of which is determined by n'. For the valve port area/travel characteristic $n' \approx 3/2$ and Grief's results show that $C_{n'}$ is given by

$$C_{n'} = \frac{2}{\sqrt{\pi}}\left\{\frac{\Gamma\left(\dfrac{n'+2}{2}\right)}{\Gamma\left(\dfrac{n'+3}{2}\right)}\right\}$$

Since $C_{n'}$ is the ratio of two gamma functions it may be evaluated from tables[27]. For $n' = 3/2$ this gives $C_{n'} = 0.914$.
The output, y, from the system is then given by

$$\frac{dy}{dt} = E^{3/2}C_{3/2}\sin(\omega t)$$

which on integration gives

$$y = \{-E^{3/2}C_{3/2}\cos(\omega t)\}/\omega \tag{7.39}$$

so that the amplitude ratio and the phase angle ϕ are given by

$$\frac{y}{x} = \frac{E^{3/2}C_{3/2}}{A'\omega} \tag{7.40}$$

and

$$\phi = \frac{\pi}{2} + \beta' \tag{7.41}$$

Substituting equations 7.37, 7.38 and 7.39 into the equation

$$\varepsilon = (x - y)$$

and equating sine and cosine terms then gives

$$E - A'\cos\beta' = 0 \tag{7.42}$$

and

$$\frac{E^{3/2}C_{3/2}}{\omega} + A'\sin\beta' = 0 \tag{7.43}$$

which on eliminating β' gives

$$\frac{A'}{\omega^2} = \frac{1-(E/A')^2}{(C_{3/2})^2(E/A')^3} \tag{7.44}$$

and from equations 7.42 and 7.43 the phase angle ϕ is given by

$$\phi = -\sin^{-1}(E/A') \tag{7.45}$$

Considering equation 7.44 it can be seen that if a value of (E/A') is chosen then the corresponding value of A'/ω^2 can be calculated and this relationship is shown in *Figure 7.28a* for $C_{3/2} = 0.914$. The reciprocal of the square root of A'/ω^2 is $\omega/(A')^{1/2}$ which can be

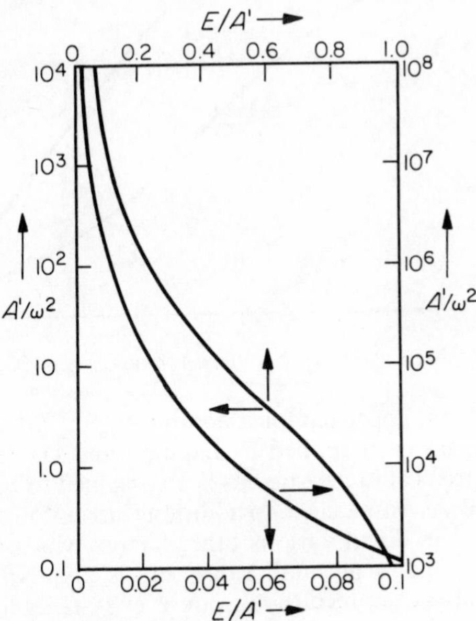

Figure 7.28(a). Variation of A'/ω^2 with ratio E/A'

regarded as a frequency and input parameter and the amplitude ratio y/x is given by equation 7.23 which may be written as

$$\frac{y}{x} = \left(\frac{E}{A}\right)^{3/2}\left(\frac{A'}{\omega^2}\right)^{1/2} C_{3/2} \tag{7.46}$$

so that the amplitude and phase angle can then be derived. (A more direct evaluation of the amplitude ration can be obtained from equations 7.40 and 7.44 i.e. $y/x = \sqrt{\{1-(E/A')\}}$ but there is then no frequency term present and results cannot be shown as a function of ω.) The variations of these two quantities with the input amplitude and frequency parameter $\omega/(A')^{1/2}$ are also shown in *Figure 7.28b* together with two exact solutions obtained from a computer.

Defining the non-dimensional 'break frequency', $\bar{\omega}_B$, as that at which the amplitude ratio equals $1/\sqrt{2}$ gives from equations 7.44 and 7.46

$$\bar{\omega}_B = 0.769(A')^{1/2}$$

or

$$\omega_B T_c' = 0.769(x/R)^{1/2} \tag{7.47}$$

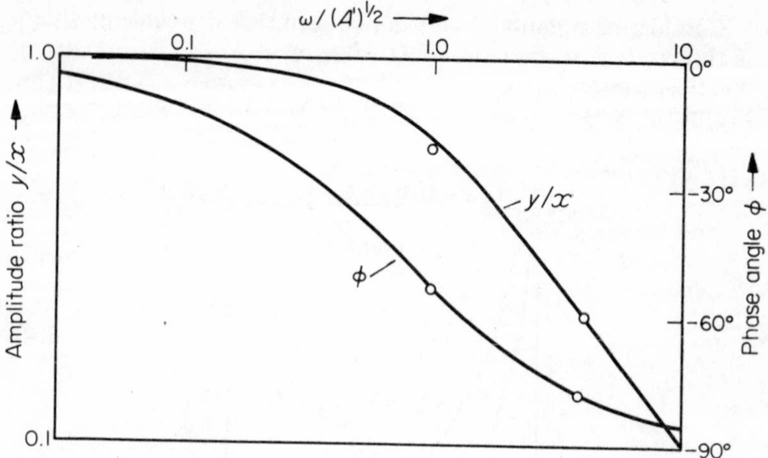

Figure 7.28(b). Variation of response with $\omega/(A')^{\frac{1}{2}}$ × exact solutions from a computer

A more general approach for a non-linear port area/spool travel characteristic has been applied by Lambert and Davies[28] in which a valve port area characteristic has been assumed to be of the form $A_p \propto z^n$ where n is unspecified. In addition Carrington and Martin[29] have discussed in detail various other factors which influence the behaviour of systems at small valve travels and Nikifarouk and Westlund[30] have examined the response of systems to large input signals.

7.6.7 Reaction forces

During the derivation of the response equation it was assumed that the valve travel z was given by the simple equation

$$z = (x - y)$$

In practice, however, the error of the system $(x - y)$ will often pass through the components shown in *Figure 6.1* (Chapter 6) and, as a result, the error will be amplified by a factor K_A where K_A is the gain of the amplifier. This signal will be applied to the transducer or torque motor which in turn will produce a force F_t given by

$$F_t = K_t K_A (x - y) \tag{7.48}$$

where K_t is the torque motor constant.

The torque motor force F_t will act on the valve spool so that its position z will be given by

$$F_t = F + m\ddot{z} + k_v \dot{z} + sz$$

where F is the total reaction force, m is the spool mass, k_v is its viscous friction coefficient (see Reference 13) and s is the rate of the spring attached to the spool opposing its motion. In practice the acceleration term, $m\ddot{z}$, and the viscous damping term, $k_v \dot{z}$, are usually small and may therefore generally be neglected, giving

$$F_t = F + sz \tag{7.49}$$

Substituting for the reaction force F from equation

$$F_t = KP_v z + sz = z(KP_v + s)$$

and combining this with equation 7.48 above and rearranging them gives

$$z = \frac{K_t K_A (x-y)}{\{s + KP_v\}} \tag{7.50}$$

If this equation is used in the derivation of the response equation and the KP_v term is written as $K(P_s - P_r)$ with $P_r = M\ddot{y}/A_r$ then

$$\dot{y} = \frac{(x-y)K_t K_A \sqrt{(1 - M\ddot{y}/P_s A_r)}}{sT\left\{1 + \dfrac{KP_s}{s}(1 - M\ddot{y}/P_s A_r)\right\}}$$

Defining a new time constant T_F to include the term $K_t K_A/s$ such that

$$T_F = Ts/K_t K_A \tag{7.51}$$

and spring rate ratio λ given by

$$\lambda = KP_s/s \tag{7.52}$$

and using a similar notation to that used when deriving equation 7.7 then gives

$$\dot{\bar{y}} = \frac{(1-\bar{y})\sqrt{(1 - L_m \ddot{\bar{y}})}}{1 + \lambda + \lambda L_m \ddot{\bar{y}}} \tag{7.53}$$

The physical significance of the spring rate ratio λ is that it represents the ratio of the equivalent spring rate of the force reaction, when the whole of the supply pressure P_s acts across the control ports, to the rate of the spring attached to the spool.

7.6.8 The response when reaction forces are present

If reaction forces are present but the inertial load is small then $L_m \to 0$ and the response equation becomes,

$$\dot{\bar{y}} = \frac{(1-\bar{y})}{(1+\lambda)}$$

and in effect this is equivalent to an increase in the time constant of the system by a factor of $(1+\lambda)$. This condition has been studied by Williams[31]. Although he stressed that the reaction forces produced a reduction in gain (equivalent to the increase in the time constant mentioned above) this is hardly a fair comment since in practice one should reduce the spring rate of the spring attached to the spool by an equivalent amount (KP_s) before comparing results. If this is done then the system has a certain useful property insofar as the flow rate/valve pressure drop characteristic tends to become flattened ($C_p \to 0$) and the valve becomes an almost constant flow rate device. This is because the input to the valve is no longer a displacement but a force, which is balanced by the spring and reaction forces. If a given signal is applied to the system and if after the valve has opened the valve pressure drop tends to decrease then the reaction force will decrease and the valve will open further until a new force balance condition is reached. In other words since the flow rate is given by

$$Q \propto z\sqrt{P_v}$$

and from equation 7.50

$$z \propto (x-y)/(s+KP_v)$$

the condition may be written as

$$Q \propto \frac{(x-y)\sqrt{(P_v)}}{(s+KP_v)} \tag{7.54a}$$

and any reduction in the flow rate caused by a reduction in the P_v term in the numerator will be less since there will be a corresponding reduction in the magnitude of the denominator.

This is an important feature of systems having valves with significant reaction forces and was probably first reported in

Reference 25 from which the curves in *Figure 7.29* are taken. Since then, further studies have been made by McCloy and Martin[32]. The curves shown in *Figure 7.29* are presented in the form of a flow rate ratio q/Q and a supply pressure ratio p_s/P_s where the flow rate Q and the supply pressure P_s are those which exist when reaction forces

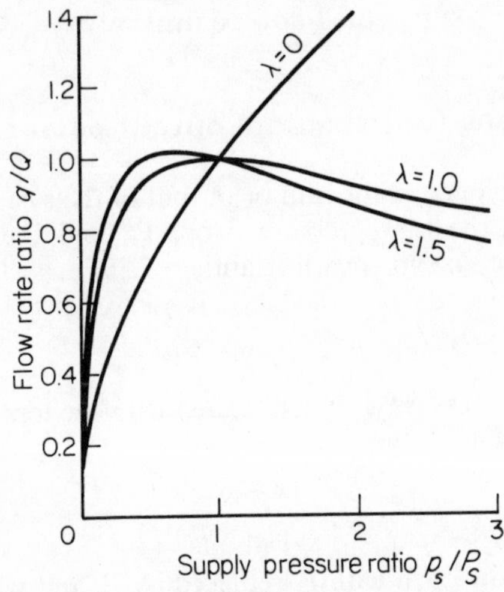

Figure 7.29. Variation of flow rate ratio with supply pressure ratio when reaction forces are present

are absent and the spring ratio then being zero. If there is no load on the system then $P_v = P_s$ and equation 7.54a becomes

$$Q \propto \frac{(x-y)\sqrt{(P_s)}}{s(1+\lambda' P_s)} \tag{7.54b}$$

where $\lambda' = \lambda/P_s$

and if Q and P_s change to q and p_s this gives

$$q \propto \frac{(x-y)\sqrt{p_s}}{s(1+\lambda' p_s)}$$

If $Q = P_s = 1$ then the flow rate ratio q/Q is given by

$$\frac{q}{Q} = \frac{(1+\lambda)\sqrt{(p_s/P_s)}}{(1+\lambda p_s/P_s)} \tag{7.54c}$$

and this is the equation used in plotting the curves shown in *Figure 7.29*. It should be noted that when $\lambda = 1.0$ the system is remarkably insensitive to supply pressure changes over the range

$0.4P_s$ to $2.4P_s$ so that a change in the supply pressure over a six to one range would have little effect on the system's behaviour in terms of flow rate. The converse of this is also applicable since the term p_s/P_s may also be thought of as $P_v/P_s = (1 - P_r/P_s)$ and it can be seen that load pressure drop (P_r) will have little effect on the flow rate over the same range. The valve may therefore be thought of as a pressure compensated flow control valve.

7.7 Conditions for maximum output power

The power developed at the ram or output of the system is given by the product of the ram pressure drop, P_r, and the flow rate Q. Considering the system shown in *Figure 7.30*, these two quantities are given by

$$P_r = P_s - P_v - P_L \tag{7.55}$$

where P_L is the pressure drop associated with the losses in the pipework of the system

and
$$Q \propto \frac{(x-y)\sqrt{P_v}}{(1+\lambda' P_v)} \tag{7.56}$$

which is equation 7.37b with P_s replaced by P_v and which gives the flow rate for a given system error $(x-y)$.

The pressure loss, P_L, in the pipework may be written in the form

$$P_L = k_L Q^{m'} \tag{7.57}$$

where k_L is a loss coefficient and m' is a loss index which if turbulent flow conditions exist is generally approximately equal to 2.0.

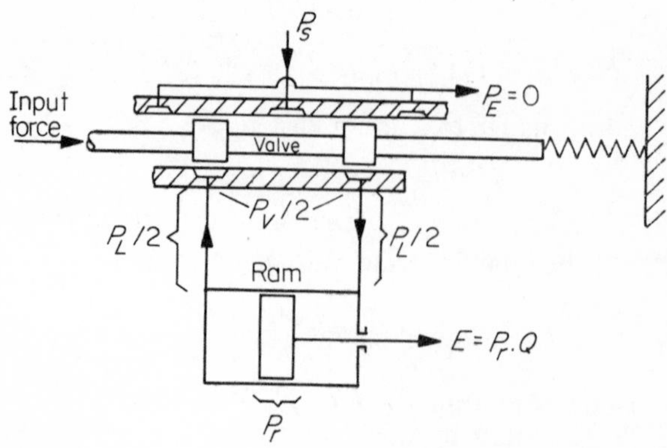

Figure 7.30. Conditions for maximum system output power

The output power available at the ram, E, is then given by $P_r Q$ and this may be written in terms of P_s and P_v by substituting equations 7.55, 7.56 and 7.57 into the equation

$$E = P_r Q$$

i.e.

$$E \propto \left[\frac{P_s \sqrt{P_v}}{(1+\lambda' P_v)} - \frac{P_v}{(1+\lambda' P_v)} - \frac{k_L (P_v)^{\{(m'+1)/2\}}}{(1+\lambda' P_v)^{\{m'+1\}}} \right]$$

Differentiating the above expression with respect to P_v and equating to zero for a maximum finally gives

$$P_s = \frac{3P_v}{(1-\lambda' P_v)} \left[1 + \frac{\lambda' P_v}{3} + \frac{k_L (m'+1)(1+\lambda' P_v)^{\{m'+1\}} P_v^{\{(m'-2)/2\}}}{3} - \frac{2k_L \lambda' (m'+1) P_v^{m'}}{3(1+\lambda' P_v)} \right] \quad (7.58a)$$

which relates the value of P_s to that of P_v, giving the maximum value of the output power.

Three simpler cases are also of interest

$$\lambda' = k_L = 0$$
$$\lambda' = 0; \quad k_L \neq 0$$
$$\lambda' \neq 0; \quad k_L = 0$$

and these give the following results:

if $\lambda' = k_L = 0$; then

$$P_s = 3P_v \quad \text{or} \quad P_v = P_s/3 \quad (7.58b)$$

if $\lambda' = 0$, $k_L \neq 0$;

$$P_s = P_v [3 + k_L (m'+1) P_v^{\{(m'-2)/2\}}] \quad (7.58c)$$

and if $\lambda' \neq 0$, $k = 0$;

$$P_s = P_v (3 + \lambda' P_v)/(1 - \lambda' P_v) \quad (7.58d)$$

Equations 7.58 may be used to ensure that the maximum value of the output power is utilised for any system.

REFERENCES

1. Coombes, J. E. M., 'Hydraulic remote position-controllers', *J.I.E.E.* **94,** Pt. 2a, 2, p. 270 (1947).
2. Grosser, C. E., 'Valve controlled servomechanisms', *Applied Hydraulics* **1,** 7, p. 15; 8, p. 16 (1948).
3. Gold, H., Otto, E. W., and Ransom, V. L., 'An analysis of the dynamics of hydraulic servomotors under inertia loads and the application to design', *Trans. ASME* **75,** p. 138 (1953).
4. Lewis, C. W., 'Some factors influencing the speed of response of hydraulic position servomechanisms', *R.A.E.*, T.N., G.W.378 (1955).
5. Heinz, R. P., 'A new maximum power criterion for certain hydraulic control valves', *ASME*, Paper No. 56-A-158B (1956).
6. Conway, H. G. and Collinson, E. G., 'An introduction to hydraulic servomechanism theory', *Proc. Conf. Hyd. Servos. Mech.E.*, p. 1 (1953).
7. Cook, A. E. and Heaps, H. S., 'Series analysis of a closed-loop system containing a loaded hydraulic relay'. *Proc. I.Mech.E.* **179,** Pt I (1964/65).
8. Ruggles, R., 'Failure-survival automatic flight control systems for aircraft with particular reference to a high reliability electrohydraulic actuator', *Proc. I.Mech.E.* **180,** Pt I (1965/66).
9. Foreman, P. F. and Carrington, J. E., 'Servo reliability in the *Seacat* missile', *Proc. I.Mech.E.* **180,** Pt I (1965/66).
10. Mucha, E., 'Transient response of hydraulic actuators', *Prod. Eng'g.* **28,** 4, p. 175 (1957).
11. Stallard, D. V., 'Single vane hydraulic servomotors and a hydraulic damper', *ASME Paper No. 57-A-127* (1957).
12. Royle, J. K., 'Hydraulic damping techniques at low velocity', *Paper 10., Proc. Conf. Oil Hyd. Power Trans. and Control, I.Mech.E.* (1961).
13. Noton, G. J. and Turnbull, D. E., 'Some factors influencing the stability of piston-type control valves', *Proc. I.Mech.E.* **172,** p. 1065 (1958).
14. Royle, J. K., 'Inherent non-linear effects in hydraulic control systems with inertia loading', *Proc. I.Mech.E.* **173,** p. 257 (1959).
15. Khoklov, B. A., *Electrohydraulically controlled drives.* 1st edition, Hayka, Moscow (1964).
16. Turnbull, D. E., 'The response of a loaded hydraulic servomechanism', *Proc. I.Mech.E.* **173,** p. 270 (1959).

17a. Diprose, K. V., *Automatic and manual control*, 1st edition, p. 304, Butterworth, London (1952).

17b. McCloy, D., 'Graphical analysis of the step response of hydraulic servomechanisms', *Proc. I.Mech.E.* 182, Pt 1 (1967/68).

18. Huddlestone, F. J., 'Graphical analysis of hydraulic servos', *Cont. Eng'g.* **5,** 4, p. 89 (1958).
19. Walters, R., 'Hydraulic and electro/hydraulic servo systems', 1st edition, p. 51, Iliffe, London (1967).
20. Harpur, N. F., 'Some design considerations of hydraulic servos of jack type', *Proc. Conf. Hyd. Servos. I.Mech.E.*, p. 41 (1953).
21. Faisandier, J., *Les Mècanismes Hydrauliques*, 1st edition, Dunod, Paris (1957).
22. Friedrich, H. R., 'A method for investigating the influence of flexibility of the mounting structure of hydraulic servo systems on the dynamic stability quality of control systems', *J. Aero. Sci.* **22,** p. 101 (1955).
23. Houlobek, F., 'Impedance of idealised hydraulic valve controlled servos', *R.A.E.* TN., ME, 108 (Dec. 1951).

24. Watson, C. D., 'Variable hydraulic pump servos, and a method of testing', *Conf. Hyd. Servos. I.Mech.E.*, p. 12 (1953).
25. Turnbull, D. E., *Piston-type control valves*, Ph.D. Thesis, Cambridge (1956).
26. Grief, H. D., 'Describing function method of servomechanism analysis applied to most commonly encountered nonlinearities', *Appl. and Ind.* p. 243 (Sept. 1953).
27. Flügge, W., *Four place transcendental functions*, Pergamon Press, London (1954).
28. Lambert, T. H. and Davies, R. M., 'Investigation of the response of an hydraulic servomechanism with inertial load', *Jnl. Mech. Eng'g. Sci.* **5,** No. 3, p. 281 (Sept. 1963).
29. Carrington, J. E. and Martin, H. R., 'Threshold problems in electrohydraulic servomotors', *Proc. I.Mech.E.* **180,** Pt I (1965/66)
30. Nikiforuk, P. N. and Westlund, D. R., 'The large signal response of a loaded high-pressure hydraulic servomechanism'. *Proc. I.Mech.E.* **180,** Pt I (1965/66).
31. Williams, H., 'The effect of oil momentum forces on the performance of electro-hydraulic servomechanisms', *Proc. Symp.—Recent mechanical engineering developments in automatic control, I.Mech.E.* (1960).
32. McCloy, D. and Martin, H. R., 'Some effects of cavitation and flow forces in the electrohydraulic servomechanism', *Proc. I.Mech.E.* **178,** Pt I, p. 539–58 (1964).

Index

Note. Numbers in bold type indicate complete chapters